Chemistry and Pharmacology of Natural Products

T0275709

Plant lectins

CHEMISTRY AND PHARMACOLOGY OF NATURAL PRODUCTS

Series Editors: Professor J.D. Phillipson, *Department of Pharmacognosy, The School of Pharmacy, University of London;* Dr D.C. Ayres, *Department of Chemistry, Queen Mary College, University of London;* H. Baxter, *formerly at the Laboratory of the Government Chemist, London.*

Also in this series
Edwin Haslam *Plant polyphenols: vegetable tannins revisited*
D.C. Ayres & J.D. Loike *Lignans: chemical, biological and clinical properties*

Plant lectins

A. PUSZTAI

The Rowett Research Institute, Aberdeen

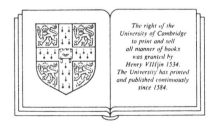

The right of the
University of Cambridge
to print and sell
all manner of books
was granted by
Henry VIII in 1534.
The University has printed
and published continuously
since 1584.

CAMBRIDGE UNIVERSITY PRESS

Cambridge

New York   Port Chester

Melbourne   Sydney

CAMBRIDGE UNIVERSITY PRESS
Cambridge, New York, Melbourne, Madrid, Cape Town, Singapore, São Paulo

Cambridge University Press
The Edinburgh Building, Cambridge CB2 2RU, UK

Published in the United States of America by Cambridge University Press, New York

www.cambridge.org
Information on this title: www.cambridge.org/9780521328241

First published 1991

A catalogue record for this publication is available from the British Library

ISBN-13 978-0-521-32824-1 hardback
ISBN-10 0-521-32824-1 hardback

Transferred to digital printing 2006

Contents

Acknowledgements

The author is a Senior Research Fellow of the Rowett Research Institute. Both his research carried out over the years and writing of this book have been made possible by Institute's generous support.

The author is also indebted to The Leverhulme Trust for the award of an Emeritus Fellowship and to the Royal Society of Edinburgh for the Auber Bequest Award. Their support in the final phases of this work is gratefully acknowledged. The author sincerely thanks the many colleagues and, especially, Drs S. Bardocz and K. Baintner, for the valuable discussions, their time, substantial help and generous advice with the manuscript during the compilation and checking of data and the writing of the book. The micrographs in Figures 6.2, 6.3, 6.4, 6.13 and 6.14 were taken by Dr T.P. King and his permission to include them in this book is much appreciated. Our work is a part of a FLAIR Concerted Action Programme (No. 9) supported by the Commission of European Communities and coordinated by the author.

Introduction

It is just over a hundred years ago that the first plant lectin was described by Stillmark (Stillmark, 1888, 1889). Working with extracts from castor bean (*Ricinus communis*) he obtained a preparation which agglutinated red blood cells. As more and more of such substances were later discovered in other plants, and as a common name for them, the term, haemagglutinin, was proposed by Elfstrand (1898). The striking toxicity of some of the haemagglutinins, such as ricin and abrin (*Abrus precatorius*) has made the effects of these substances relatively easy to test on animals. Ehrlich, who is usually considered to be the father of immunology, has shown that rabbits fed with small amounts of seeds containing the toxin became partially immune to the toxicity, thus demonstrating that the haemagglutinins were also antigenic. Landsteiner and Raubitschek (1908) showed later that not all haemagglutinins need necessarily be as toxic as ricin or abrin. For example, the agglutinins obtained from common beans (*Phaseolus vulgaris*), peas (*Pisum sativum*), lentils (*Lens culinaris*), etc, were relatively non-toxic, water soluble proteins. It is now known that such haemagglutinating proteins are found in all taxonomic groups of the Plant Kingdom and that they are not all overtly toxic.

The next momentous step in the history of haemagglutinins was the realization that some of the haemagglutinins agglutinated blood cells only from some groups of individuals within the ABO blood group system, without affecting cells from other groups (Renkonnen, 1948; Boyd & Reguera, 1949). Indeed, this discovery of blood group specificity has led Boyd to coin the term, lectin, to denote this aspect of selection (in Latin, *legere* means to select) and is regarded as the starting point of modern lectinology.

With our increased understanding of the chemical structure of blood group-specific glycoconjugates, the involvement of sugars in the agglutination reaction was soon appreciated. As early as 1936, Sumner and Howell observed that cane sugar inhibited the haemagglutination activity of concanavalin A (Sumner & Howell, 1936). It was, however, somewhat later that Watkins and Morgan (1952) laid the foundations of our knowledge of the strict sugar specificity of the agglutination reaction within the human ABO blood group system. Indeed, the first proper definition of lectins was based on the sugar specificity of the inhibition of the haemagglutination reaction. Accordingly, lectins are carbohydrate-binding proteins of non-immune origin which agglutinate cells or precipitate polysaccharides or glycoconjugates (Goldstein *et al.*, 1980). This definition was adopted by the Nomenclature Committee of the International Union of Biochemistry (Dixon, 1981).

The main problem with this definition is that, if it is strictly interpreted, some poorly agglutinating well-known toxins, such as ricin, abrin, modeccin, etc, cannot be regarded as lectins, even though they are all known to contain lectinic subunits. Thus, the first definition has since been extended to include the above toxins (Kocourek & Horejsi, 1983). Moreover, as it is now realized that some lectins contain a second type of binding site that interacts with non-carbohydrate ligands (Barondes, 1988), confining the definition of lectins strictly to bivalent carbohydrate-binding proteins seems to have lost its usefulness. In fact, a narrow definition may even impede our understanding of the proper endogenous function(s) of lectins. Although the more general definition of lectins (Kocourek & Horejsi, 1983) is not yet universally accepted, it is very appealing and serves as a basis for the definition of lectins in this book. Accordingly, lectins are proteins (or glycoproteins) of non-immunoglobulin nature capable of specific recognition of, and reversible binding to, carbohydrate moieties of complex glycoconjugates without altering the covalent structure of any of the recognized glycosyl ligands. Thus, other sugar-binding proteins, such as the various sugar-specific enzymes, hormones and transport proteins are excluded, but monovalent lectins (i.e. bacterial and plant toxins) are included.

1

Lectins and their specificity

The carbohydrate specificity of lectins is established most conveniently by the Landsteiner hapten-inhibition technique (Landsteiner, 1962; Goldstein & Hayes, 1978; Goldstein & Poretz, 1986). In this method the inhibitory effectiveness of various mono- and oligosaccharides of known composition is compared in a recognized and convenient reaction of lectins, such as haemagglutination. Initially, on the basis of their reactivity with monosaccharides that differed in configuration at C-3 or C-4 of the pyranose ring, Mäkelä (1957) suggested that lectins could be divided into four classes. Thus, for example, concanavalin A, which reacts with D-mannose and/or D-glucose, belongs to group III. Soyabean lectin, whose specificity is for N-acetyl-D-galactosamine and/or D-galactose, is classified as group II. According to Mäkelä's scheme the L-fucose-binding lectins are members of group I (Fig. 1.1). However, no group IV-specific lectins have been found in Nature so far. In more recent studies, although still within the four basic classes of Mäkelä, the definition of the specificity of lectins has been further refined and extended. Accordingly, from their reaction with simple sugars, lectins are first classified into broad groups of lectin classes as either mannose/glucose-specific, or N-acetylglucosamine-specific, or N-acetylgalactosamine/galactose-specific or fucose-specific lectins. This is then followed by a more precise classification based on extensive investigations with a great number of oligosaccharides of known composition and structure. From such studies of the most complementary carbohydrate structure, the recognition and binding site of the lectins within the four main classes, is obtained.

Lectins are known to react chiefly with the non-reducing end of oligo- and polysaccharides. Although there are a number of known exceptions to this rule, the specificity of lectins for the terminal sugar tolerates little

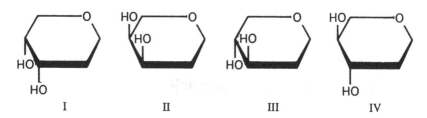

Fig. 1.1. Mäkelä's four major classes of lectins based on the configuration of hydroxyl groups at C-3 or C-4 of the pyranose ring.

variation at C-3 or C-4 of the pyranose ring. The configuration at the second carbon atom, however, appears to be less critical.

In view of their strict carbohydrate specificity, it is not surprising that most lectins have been purified to a high degree of purity by affinity methods. Thus, although some lectins, particularly those with unknown or unusual sugar specificities, in the past, have also been purified to homogeneity by the application of conventional protein purification methods, isolation based on affinity chromatography (Goldstein & Poretz, 1986) is now almost the rule. Indeed, in most instances, homogeneous lectin preparations can easily be obtained, in one-step procedures, by chromatography on affinity columns which contain insolubilized specific haptens bound to suitable supports.

D-mannose/D-glucose-specific lectins

This group contains the best known and most studied lectin: concanavalin A from jack bean (*Canavalia ensiformis*). Its structure, however, is somewhat atypical as concanavalin A is composed of four identical subunits. In contrast, most other lectins in the group, such as those obtained from peas (*Pisum sativum*), broad beans (*Vicia faba*) or lentils (*Lens culinaris*) consist of two light (α) and two heavy (β) chains and have a general subunit structure of $\alpha_2\beta_2$.

Purified preparations of concanavalin A were obtained as early as 1936. Indeed, this lectin was one of the earliest examples of a well-characterized crystalline protein described (Sumner & Howell, 1936). It was also one of the first lectins isolated by affinity chromatography (Agrawal & Goldstein, 1967). At or above pH 7, concanavalin A contains four non-glycosylated polypeptide chains of $M_r = 26,500$. Each subunit is made up of 237 amino acid residues. Although some of the peptide bonds in the primary sequence of the subunits are known to be broken, as the fragmented polypeptide chains are held together by non-covalent forces, the native conformation of the protein molecule is maintained regardless of whether the subunits are fragmented or not (Wang, Cunningham &

Edelman, 1971). The major breakage point in the polypeptide chain has been shown to be between Asn(118) and Ser(119) residues. When such concanavalin A preparations are dissolved in dissociating solvents, such as SDS, and electrophoresed on polyacrylamide gels in SDS, in addition to the major band of the unbroken subunit (M_r = 26,500), two smaller size polypeptide bands are also obtained. Rather interestingly, although concanavalin A is not a glycoprotein, it is, in fact, first synthesized as an inactive glycoprotein, with a leader sequence, in the developing seeds. In the inactive concanavalin A precursor, the amino terminal part of the active lectin which is characteristic for the lectin found in mature seeds, is in a midchain position. During seed development, the glycoprotein is modified by proteolytic processing and splitting at this midchain position and the leader sequence is removed from the amino terminus. The resulting two polypeptide chains are then ligated to each other in such a way that the amino terminal liberated from the middle of the precursor polypeptide chain becomes the new amino terminal of the mature form of the lectin, while the amino terminal of the precursor is joined to its carboxyl terminal residue by a new peptide bond. During this process, the carbohydrate part, which in the precursor shielded the carbohydrate-binding site of the lectin, is also removed. With this unmasking of its functional amino acid side-chains, the non-glycoprotein precursor is transformed into the fully active lectin, concanavalin A (Carrington, Auffret & Hanke, 1985) (Fig. 1.2).

Concanavalin A is a metalloprotein in which each subunit contains one Ca^{2+} and one Mn^{2+}. When the metal ions are removed, the lectin loses its activity (Agrawal & Goldstein, 1968). Apparently, the metal ions are necessary to lock the conformation of the lectin in a form in which the carbohydrate-binding sites are correctly exposed (Strazza & Sherry, 1982). Concanavalin A was also the first lectin whose complete three-dimensional structure has been determined by X-ray diffraction methods (Edelman *et al.*, 1972). Accordingly, the subunits of the lectin are compactly folded and dome-shaped structures. In the native concanavalin A, two of the subunit polypeptides with a structure based on antiparallel β-pleated sheets are joined in a functional ellipsoidal dimeric unit and two of these are then paired to form the native tetrameric lectin (Fig. 1.3).

The various functional sites in the lectin molecule have also been established. Thus, for example, the metal-binding sites are known to be in the amino terminal sequence of the lectin. The carbohydrate-binding site was determined on crystals obtained from the carbohydrate complexes of concanavalin A. Several amino acids are implicated in sugar

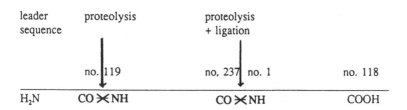

Fig. 1.2. Post-translational changes in the primary amino acid sequence of concanavalin A during seed maturation and homologies to other Vicieae lectins.

binding. Although these are in widely different positions in the primary amino acid sequence of the subunit polypeptides, spatially they are closely located in the native conformation of the lectin molecule. The amino acid residues involved are: Tyr(12), Asn(14), Asp(16), Leu(99), Tyr(100), Ser(168), Asp(208) and Arg(228).

In the classical early studies of Goldstein and his associates (for refs, see Goldstein & Poretz, 1986) it has been established that, of the monosaccharides, mannose, in its α-anomeric form, is the most active simple sugar inhibitor of the biological activity of concanavalin A. Moreover, in oligo- and polysaccharides, first and foremost, it is the non-

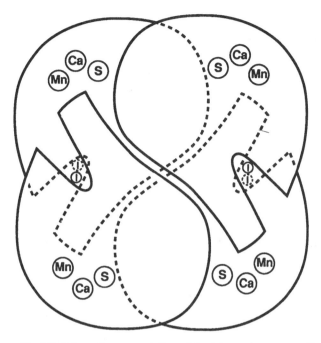

Fig. 1.3. Schematic representation of the tetrameric concanavalin A molecule with its metal (Ca^{2+} and Mn^{2+}), carbohydrate (S) and hydrophobic (I) binding sites (from Becker *et al.*, 1976; reproduced by permission of the authors and *Nature*).

reducing terminal mannose residue which is recognized by the lectin. Although this appears to be generally true for most polysaccharides interacting with concanavalin A, in α-(1–2) linked mannose-containing oligosaccharides, other mannose residues, and not just the terminal non-reducing ones, make appreciable contributions to the interaction. The results have, in fact, suggested that the best fit for the concanavalin A-carbohydrate-binding site is likely to be oligosaccharides containing three to four α-(1–2) mannosyl residues. Concanavalin A binds to branched trimannosyl structures, present in ovalbumin glycopeptides, with great affinity. These show similarity to the synthetic 3,6-di-*o*-(α-mannosyl)-mannose (Brewer *et al.*, 1985). Moreover, concanavalin A can be precipitated with this high mannose glycopeptide (D_3) from ovalbumin. Thus, one lectin reactive site, a trimannosyl moiety, is found on the α-(1–6) arm of the core β-mannose residue. Another site is found on the α-(1–3) arm of the core containing an α-(1–2) mannobiosyl group. Similar high mannose-type glycopeptides on the surface of a number of cells may also function as bivalent cross-linking agents and play a part in lectin-induced patching and capping. Thus, these surface sugars may be involved in signal transduction as specific receptors (Bhattacharyya & Brewer, 1986).

Concanavalin A contains a number of non-polar binding sites of varying affinities. A high affinity site, which is close to the carbohydrate-binding site and reacts with phenyl groups in phenylglucosides and phenylmannosides, has been described (Poretz & Goldstein, 1971). Another high affinity site interacts with fluorescent hydrophobic probes (Roberts & Goldstein, 1983a) and one low affinity site per subunit binds tryptophan and indoleacetic acid (Edelman & Wang, 1978). There is also a non-polar binding cavity. This, however, is present only in concanavalin A crystals. It is clear that, with suitable ligands, non-polar binding sites can make considerable contributions to the primary sugar-specific binding and result in an increased and tighter overall binding between lectin and ligands. The functional importance of hydrophobic binding is clearly shown by the observation that even monomeric concanavalin A derivatives agglutinate guinea pig erythrocytes (Ishii *et al.*, 1984). The more recent demonstration of the specific binding of *myo*-inositol by concanavalin A, which is independent of the mannose/glucose binding site, further demonstrates that interactions between lectins and naturally occurring glycoconjugates are not necessarily confined to specific carbohydrate structures in the ligands (Wassef, Richardson & Alving, 1985). Thus, concanavalin A binds strongly to phosphatidylinositol-containing liposomes, even when these contain no sugar residues.

Several other lectins belonging to this class of mannose/glucose-specific haemagglutinins have been described in some detail over the years. One of these is the rather interesting close homologue of concanavalin A, the lectin isolated from the seeds of *Dioclea grandiflora* (Moreira *et al.*, 1983). Indeed, both plants from which these lectins have been isolated belong to the same tribe of Diocleae. The *Dioclea grandiflora* lectin is a metallo-protein, devoid of covalently bound carbohydrate residues and has the same specificity for D-mannose/D-glucose residues as concanavalin A. Its molecular weight and subunit structure, dissociation properties and the fragmentation patterns of its subunits resemble those of concanavalin A. Although 53 of the 237 amino acid residues of the intact α-chain polypeptide are different from those found in concanavalin A, six out of the total seven residues implicated in metal binding are conserved in the *Dioclea grandiflora* lectin (Richardson *et al.*, 1984).

Several mannose/glucose-binding lectins from the *Lathyrus* genus have been studied extensively. These lectins all appear to be very similar proteins containing no covalently bound carbohydrate residues, with M_r

values close to 50,000 and with a general subunit structure of $\alpha_2\beta_2$. Their specificity is similar to that of concanavalin A.

The class of D-mannose/D-glucose-binding lectins also contain a lectin isolated from lentils (*Lens culinaris*) with its main specificity for non-reducing α-mannopyranosyl terminal residues. This lectin also binds to oligo and polysaccharides which contain internal 2-*o*-substituted mannose residues. Glycoconjugates in which an α-L-fucosyl group is linked to *N,N'*-diacetylchitobiosyl units are also known to bind tightly to the lentil lectin. Similar lectins, with sugar-binding and molecular properties close to those of the lentil lectin, have been isolated and studied, from amongst others, peas (*Pisum sativum*), common vetch (*Vicia cracca*) and other members of the *Vicia* genus and sainfoin (*Onobrychis viciifolia*) (see Goldstein & Poretz, 1986).

N-acetyl-D-glucosamine-binding lectins

In this diverse group of lectins, whose members are found in Gramineae, Solanaceae and also in Leguminoseae, carbohydrate specificity is directed against *N*-acetyl-D-glucosamine and, particularly, its β-(1–4) linked oligomers.

The lectin from potato (*Solanum tuberosum*) tubers, a glycoprotein containing a fairly large covalently bound carbohydrate component, was first purified by conventional protein purification methods (Marinkovich, 1964). The lectin is unusual as it is readily inactivated by reduction with 2-mercaptoethanol. Purification of the lectin, however, is much more easily achieved by affinity chromatography. Indeed, several different affinity supports have been used successfully in the past for the isolation of homogeneous preparations of potato lectin. These, for example, have included fetuin-Sepharose (Owens & Northcote, 1980) or *N,N',N''*-triacetylchitotriose–Sepharose (Desai & Allen, 1979).

Potato lectin has a molecular weight of about 100,000 and contains two non-covalently linked identical subunits (Allen & Neuberger, 1973; Owens & Northcote, 1980). Each subunit of the glycoprotein lectin is made up of two dissimilar domains. One of the domains contains all the hydroxyproline and carbohydrate residues. This part of the molecule also contains about half of the total number of serine residues and these are all galactosylated. The carbohydrate part of this domain contains mainly L-arabinose. Indeed, over 90% of the sugar constituents of the carbohydrate part consists of arabinose, while the rest is galactose (Allen *et al.*, 1978). In fact, the overall composition of this domain of the potato lectin resembles closely most of the structural features of plant cell wall glycoconjugates. The bulk of the arabinose occurs in short linear chains of three or four arabinofuranosyl groups linked to hydroxyproline (Allen *et al.*, 1978; Ashford *et al.*, 1982). All linkages in the triarabinoside are β-(1–2) and in the tetraarabinoside the last residue is linked by an α-(1–3) linkage (Ashford *et al.*, 1982).

The other domain of the lectin subunit is very different and contains all the cysteine residues of the molecule. The carbohydrate-binding site of the lectin is also located in this domain. Not surprisingly, due to the substantial differences in the two domains, antibodies raised against the intact potato lectin contain two totally different populations of immuno-globulins. One population reacts exclusively with the glycosylated part of the molecule, while the cysteine-rich part of the lectin is recognized only by the second immunoglobulin component of the antibody mixture. The fundamental difference between the two domains is also shown by the observation that antibodies against the glycosylated domain cannot abolish the biological activity of potato lectin, while reaction of the intact lectin with the antibody population against the cysteine-rich part strongly inhibits haemagglutination activity. Similarly, most of the covalently bound sugar residues can be removed from the lectin by treatment with trifluoromethanesulphonic acid, without affecting its activity. Accordingly, the active site of potato lectin containing the carbohydrate recognition and binding sites is confined to the cysteine-rich and non-glycosylated domain of the lectin molecule.

Oligosaccharides of N-acetyl-D-glucosamine are all good inhibitors of the haemagglutination activity of potato lectin (Allen & Neuberger, 1973). However, N-acetyl-D-glucosamine itself is very poorly inhibitory. In fact, the inhibitory potency of the monosaccharide is of six orders of magnitude less than that of N,N'-diacetylchitobiose. Moreover, in N-peracetylated chitooligosaccharides, the inhibitory activity increases with increasing chain length. Thus, N,N',N'',N''',N''''-pentaacetylchito-pentaose is about 50 times more effective in inhibiting the haemagglutination activity of potato lectin than N,N'-diacetylchitobiose. These results indicate that the sugar-complementary site of potato lectin is fairly extensive in size. With equilibrium dialysis technique Ashford *et al.* (1982) showed that each subunit ($M_r = 50,000$) has only one reactive site. Thus, for high haemagglutination potency of the native dimer lectin, both subunits take part in carbohydrate binding.

A lectin, similar to the extensively studied tuber lectin, has also been isolated from the pericarp of the potato fruit (Kilpatrick, 1980*b*). This lectin is very similar to the tuber lectin in both molecular properties and activity. In fact, all lectins from potatoes (and even from *Datura* species) are known to show extensive immunochemical cross-reactivity. However, in contrast to the behaviour of the tuber lectin, acetylchitobiose is a poor inhibitor of the haemagglutination activity of the lectin isolated from the seeds of the potato plant.

Another important lectin of the *N*-acetyl-D-glucosamine-binding class of lectins is found in the wheatgerm. This lectin was first observed as a contamination in wheatgerm lipase preparations (Aub, Tieslau & Lankester, 1963). It has subsequently been purified and crystallized (LeVine, Kaplan & Greenaway, 1972; Nagata & Burger, 1972; Allen, Neuberger & Sharon, 1973). From its composition, specificity and properties (Allen *et al.*, 1973; Nagata & Burger, 1974; Lotan, Sharon & Mirelman, 1975*a*), it is known that most preparations of wheatgerm lectin may contain three to four isolectins, with distinct electrophoretic mobilities (Allen *et al.*, 1973).

The native lectin is a dimer containing two identical subunits which are not glycosylated. Each subunit contains 171 amino acid residues and has a M_r value of 21,600 (Wright, 1984). The lectin has been examined extensively by X-ray crystallography (Wright, 1981). The three-dimensional structure of the wheatgerm agglutinin protomer is made up of four homologous but spatially distinct domains (A, B, C and D). Three of these domains contain 43 and the fourth, 42 amino acid residues, with extensive homologies in their primary amino acid sequences (Wright, Gavilanes & Peterson, 1984). The four domains are held together by four disulphide bonds (Wright, 1981) and are probably the result of gene quadruplication and subsequent divergence during evolution (Wright, Brooks & Wright, 1985).

In further extensive studies, the differences in the primary sequences of isolectins WGA1 and WGA2 have been established (Wright & Olafsdottir, 1986). In addition to the replacement of two histidine residues in the B domain of WGA2 by one glutamine and one tyrosine residue, in the corresponding domain of WGA1 there are two more substitutions, one in the B and another in the C domain of the two isolectins. The degree of homology between the entire amino acid sequences of the two isolectins is estimated to be 97.3%. When the above four differences in the primary sequences are taken into account, the internal homology between the four domains is more extensive in WGA1 than in WGA2. However, when the evolutionary distance of each isolectin from the proposed ancestral sequence (Wright *et al.*, 1985) is estimated, the nucleotide base changes required for accounting for the differences between the two isolectins are not significant. This suggests a rather recent evolutionary divergence (Wright & Olafsdottir, 1986).

Each monomeric subunit has been shown to contain two distinct carbohydrate-binding sites for N-acetylglucosamine (Nagata & Burger, 1974; Wright, 1984) or its β-(1–4) linked oligomers. The best inhibitors of the haemagglutinating activity of the wheatgerm agglutinin are found in the various members of a group of N-peracetylated chitin oligosaccharides. The best fit is obtained with a molecule containing three β-(1–4) linked N-acetylglucosamine units (Allen *et al.*, 1973). In addition, however, wheatgerm lectin has a distinct affinity for N-acetylneuraminic acid (Greenaway & LeVine, 1973). Thus, cells treated with sialidase preparations react less strongly with wheatgerm agglutinin than the original cells. Moreover, the lectin has been shown to react with N-acetylneuraminic acid, its methylester and α- and β-methylketosides and sialyllactose. The binding affinity of sialic acid is less than a quarter of that of N-acetyl-D-glucosamine, N-glycolylneuraminic acid does not bind at all.

X-ray crystallography of wheatgerm lectin with various sugar derivatives of N-acetyl-D-glucosamine and N-acetylneuraminic acid (Wright, 1980*a,b*, 1984; Wright *et al.*, 1985) has confirmed the existence of two binding sites per monomer and four for the biologically active dimer of wheatgerm agglutinin (Fig. 1.4). These are located at the dimer interface between opposite domains of the monomers. Of the two carbohydrate-binding sites, only one is readily accessible to both N-acetyl-D-glucosamine and N-acetylneuraminic acid derivatives. In addition to this primary binding site, there is another, a secondary site for binding carbohydrates. However, in the crystalline form of the lectin, this secondary site is not available for the binding of N-acetylneuraminic acid. Moreover, even the binding of N-acetyl-D-glucosamine to this secondary site is weak. Furthermore, as self-interactive association between two separate lectin molecules also occurs through this secondary carbohydrate-binding site, there is competition for the site between hapten sugars and other wheatgerm agglutinin molecules. Although sugar-binding to both sites proceeds in solution (Kronis & Carver, 1985), for the proper characterization by X-ray crystallography of the secondary carbohydrate-binding site, suitable co-crystals of wheatgerm agglutinin and the appropriate sialoglycopeptides are needed. Recently, such crystals, suitable for high-resolution studies have, in fact, been obtained from wheatgerm agglutinin in the presence of the T-5 tryptic peptide of the red cell receptor, glycophorin A (Wright & Kahane, 1987).

Glycoprotein lectins, similar to those isolated from potatoes, have also been obtained from thorn apple (*Datura stramonium*) and tomato (*Lycopersicon esculentum*) juice. The amino acid and sugar compositions of the two lectins are similar and also resemble those of the potato lectin. Their carbohydrate part consists of about 90% arabinose and 10% galactose units. Additionally, when tested with antibodies raised against the *Datura* lectin, all these lectins show extensive immunochemical cross-reactivity (Kilpatrick, 1980*a*). The carbohydrate-binding site of the tomato lectin is complementary to about three or four β-(1–4) linked *N*-acetyl-D-glucosamine units. The monosaccharide, however, is not inhibitory (Kilpatrick, 1980*a*; Nachbar, Oppenheim & Thomas, 1980). Similar considerations apply to the binding specificity of the *Datura* lectin (Crowley *et al.*, 1984).

The *N*-acetyl-D-glucosamine-specific class of lectins are known to contain several other well-characterized lectins. Of these, the *Griffonia*

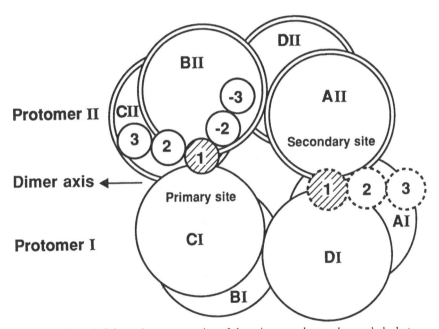

Fig. 1.4. Schematic representation of the primary and secondary carbohydrate binding sites of the dimer of wheatgerm agglutinin composed of two protomers, Protomer I and Protomer II. Each protomer consists of four domains: A, B, C and D (labelled with the appropriate subscript number). Each unique carbohydrate binding site is subdivided into subsites (small circles). The oligomers of *N*-acetyl-D-glucosamine bind at subsites of 1, 2 and 3 of both primary and secondary sites, whereas neuraminic acid-containing oligosaccharides may bind only to subsites of 1, 2 and 3 of the primary carbohydrate-binding location. Carbohydrate binding is the strongest at subsite 1 at both primary and secondary binding locations (shaded in the diagram; reproduced from Wright, 1980*a* by permission of the author and *The Journal of Molecular Biology*).

simplicifolia II lectin is well worth mentioning because, in contrast to other lectins in this group, it interacts with terminal, non-reducing α- or β-N-acetyl-D-glucosamine end groups (Iyer, Wilkinson & Goldstein, 1976). Other lectins, such as those from *Cytisus sessifolius*, *Brachypodium sylvaticum*, *Phytolacca americana*, *Ulex europeaus II*, *Wistaria floribunda*, cereals (other than wheat), members of the family Cucurbitaceae, etc, have also been described and characterized, at least partially, suggesting that this class of lectins has a very wide distribution in the Plant Kingdom (Goldstein & Poretz, 1986).

N-acetyl-D-galactosamine/D-galactose-binding lectins

Lectins which specifically recognize, and bind to, carbohydrates containing either N-acetyl-D-galactosamine or D-galactose, usually as non-reducing terminal sugar residues, have a wide distribution in Nature. Also, this class is important historically, as the first ever haemagglutinin observed and described by Stillmark (1888) and is one of the archetypes of this class of lectins. In fact, the seeds of castor bean (*Ricinus communis*, Euphorbiaceae) contain both a haemagglutinating lectin and a poorly haemagglutinating lectin-like toxin. Castor bean is not unique in this respect; other plants of diverse botanical classification are also known to contain lectins and toxins structurally related to each other and with sugar specificities characteristic for this class. Some members of this group of lectins are also blood group specific. Indeed, the lectin from lima bean (*Phaseolus lunatus*) is the first lectin described which reacts specifically with blood group A-specific cells (Boyd & Reguera, 1949). Moreover, both the lima bean lectin and the similar blood group A-specific *Dolichos biflorus* lectin show a definite preference for N-acetyl-D-galactosamine over D-galactose as the most complementary simple sugar inhibitor of haemagglutination. This is in line with the known structure of blood group A determinants, i.e. the presence of non-reducing terminal N-acetyl-D-galactosamine residues in blood group A-specific glycoproteins.

Other members of this class of lectins may be inhibited by both N-acetyl-D-galactosamine and D-galactose at concentrations of the same order of magnitude. Such lectins, for example, the lectin from *Sophora japonica*, agglutinate both blood group A or B cells and this agglutination can be inhibited by either blood group A- or B-specific substances (Morgan & Watkins, 1953).

In contrast, some other lectins in this class are specific mainly for D-galactose. For example, the haemagglutinin, RCA_I from castor bean cannot be inhibited by N-acetyl-D-galactosamine at all. Although one of the galactose-specific lectins isolated from *Griffonia simplicifolia*, GS I-

B_4, has strong preference for the agglutination of blood group B- or AB-specific erythrocytes (Goldstein & Hayes, 1978), most members of this galactose-specific lectins are not exclusively blood group B specific. Two particularly interesting lectins in this group also need mentioning. One of these has been isolated from peanuts (*Arachis hypogaea*). This lectin has anti-T, polyagglutinin activity (specificity for desialylated human red cell antigenic groups; Bird, 1954). The other lectin, from *Vicia graminea* seeds (Ottensooser & Silberschmidt, 1953), is specific for human blood group N.

Ricinus communis lectins

As mentioned before, aqueous extracts of castor bean seeds contain two different lectins. There is a strong structural resemblance between the non-cytotoxic, but powerful multivalent haemagglutinin, RCA_I, and the apparently monovalent, highly cytotoxic ricin, RCA_{II}. Both have been obtained as pure preparations by elution from Sepharose affinity chromatographic columns (Tomita *et al.*, 1972). However, as RCA_I has very little affinity for N-acetyl-D-galactosamine, the two lectins can be separated by absorption onto and sequential elution from Sepharose-4B columns. If the column is first eluted with a solution of N-acetyl-D-galactosamine, only ricin, RCA_{II}, emerges. The haemagglutinin preparation obtained with subsequent elution with D-galactose is free from the toxin, RCA_{II} (Nicolson, Blaustein & Etzler, 1974).

Ricin is a dimeric glycoprotein with a molecular weight of about 63,000. The toxin is composed of two different subunits, A and B, linked by a disulphide bridge (Nicolson *et al.*, 1974; Olsnes & Pihl, 1973, 1977). Highly purified preparations of the two separated subunits have been obtained by affinity chromatography (Olsnes & Pihl, 1973; Fulton *et al.*, 1986). The primary amino acid sequence of the two subunits has been determined (Yoshitake, Funatsu & Funatsu, 1978; Funatsu, Kimjura & Funatsu, 1979). Both chains of RCA_{II} are coded by a single mRNA (Butterworth & Lord, 1983). The primary amino acid sequence of the toxin has been deduced from the nucleotide sequence of cDNA (Lamb, Roberts & Lord, 1985). Accordingly, ricin is synthesized as a single protein, with a leader sequence on the A-chain and with a peptide containing 12 amino acids linking the A and B chains. On processing, both the leader sequence and the linking peptide are removed. The structure of the prolectin has been confirmed from the nucleotide sequence of the genomic DNA (Halling *et al.*, 1985).

Both subunits of ricin are glycosylated (Foxwell *et al.*, 1985). The sugar composition of the A polypeptide shows the presence of several mannose and *N*-acetylglucosamine residues and a single residue of both xylose and fucose. However, the A-chain is known to occur in two different forms with molecular weight values of 30,000 and 32,000, respectively. In contrast to the lighter form of the A subunit, the heavier component contains one extra carbohydrate unit. This binds to concanavalin A and is susceptible to hydrolysis by endoglycosidases. The B subunits contain either one or two high-mannose oligosaccharide side-chains. Only one of these can be removed by endoglycosidases F or H from the native molecule, while both are removable after denaturation. Intact ricin is resistant to endoglycosidase and is hydrolysed by α-mannosidase to a limited extent only (Foxwell *et al.*, 1985). The two oligomannose side-chains are also removed from the B subunit of RCA_I by treatment with preparations of the enzyme peptide: *N*-glycosidase F. However, in the presence of lactose, which binds to the lectin B subunit, only one of the sugar units was detached by the enzyme (Wawrzynczak & Thorpe, 1986).

The sugar residues of ricin can also be modified by simultaneous treatments with sodium metaperiodate and sodium cyanoborohydride, causing oxidative cleavage and subsequent reduction of the aldehydes formed to primary alcohols. All these modifications to the sugar side-chains have appreciable effects on the concentration of ricin in blood circulation. Removal of the carbohydrate side-chains of the molecule reduces the extent of the normally rapid clearance by hepatic non-parenchymal cells of ricin preparations as cytotoxins injected parenterally and targeted to kill cancerous cells (Skilleter, Price & Thorpe, 1985). Indeed, antibody-ricin conjugates prepared by recombining deglycosylated A and B ricin subunits and linking them to target-specific antibodies, are more slowly cleared and inactivated by the liver (Foxwell *et al.*, 1987).

The three-dimensional structure of RCA_{II} ricin has been determined at 2.8 Å resolution by X-ray crystallography (Montfort *et al.*, 1987). The A-chain is a globular protein with a well-ordered secondary structure and a cleft for the enzyme active site. The lectin B subunit contains two similar domains, each with a lactose-binding cleft. In the clefts, a glutamine residue is bound by hydrogen-bonding to the hydroxyl group at C-4 of the D-galactose. In addition to the known single disulphide bridge between the A and B subunits in the native molecule, strong contributions to the association between the two different polypeptides are also made by hydrophobic interactions between the various proline and phenylalanine side chains of the subunits. The location of the two glycosylation sites on the B subunit and two potential sites on the A-chain have also been indicated from the electron density maps (Montfort *et al.*, 1987).

The lectin B subunit of ricin, in contrast to previous beliefs, is now known to have two independent saccharide-binding sites. One of these is of a high affinity site, while the other has only low affinity for carbohydrates (Shimoda & Funatsu, 1985). From the reaction of the B-chain with *N*-acetylimidazole, it has been established that possibly two tyrosine residues may be involved in the sugar-reactive sites. Indeed, in the presence of lactose, the two tyrosine residues are not modified. Moreover, tyrosine-248 has been positively identified as one of such residues (Wawrzynczak *et al.*, 1987). In addition, the results of digestion studies with carboxypeptidase have suggested that the carboxyl terminal region of the B subunit does not participate directly in carbohydrate binding by the high affinity site. However, this part of the molecule is still essential for the maintenance of the active conformation and interaction with carbohydrates (Funatsu, Yamasaki & Kakinchi, 1987). As a result of the presence of two potential carbohydrate-binding sites on the B-chain, ricin possesses a weak haemagglutination activity. Moreover, these active sites on the B-chain are also responsible for binding of the toxin to cells. Thus, although the toxicity resides entirely with the A-chain, without the presence of the B-chain, ricin cannot attach itself to cells and, therefore, cannot exert its toxic effects.

Our understanding of the molecular mechanism of ricin toxicity has been developing steadily in the last few years. Accordingly, in a first step, ricin, through its B-chain with lectin activity, binds to the galactose-containing receptors found on the surface of most eukaryotic cells. This stimulates the endocytotic uptake of the entire ricin molecule by the cell (Nicolson, Lacorbiere & Hunter, 1975). Although the route from the endosomic vesicles into the cytosol is not entirely clear, there is some evidence to suggest that the B subunit makes a vital contribution to this (Simmons, Stahl & Russell, 1986) and that a proportion of the internalized lectin, about 5%, is delivered from the endosomes into the cisternae of the Golgi apparatus (van Deurs *et al.*, 1987). This part of the ricin, still bound to co-internalized plasma membranes, reaches the *trans*-Golgi network and does not end up in the lysosomes (Fig. 1.5). The results also suggest that, from the internalized ricin-plasma membrane complex, the membrane glycoproteins which contain the terminal galactose residues participating in ricin-binding are recycled from the Golgi to the cell surface and not from the endosomes or lysosomes. The ricin A-chain is then detached both from the membranes and the B subunit and translocated to the cytosol from the *trans*-Golgi network (van Deurs *et al.*, 1988), where it enzymically attacks the 60S ribosomal subunit and halts protein synthesis (Olsnes & Pihl, 1982). The enzymic attack of the ricin A-chain is

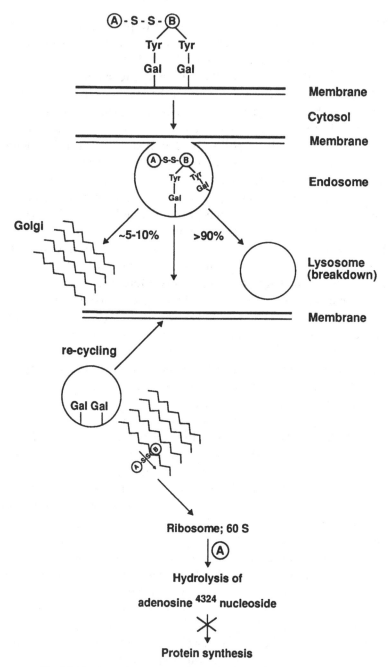

Fig. 1.5. A schematic representation of the intracellular pathways and reactions of the endocytosed intact ricin molecule leading to irreversible stoppage of cellular protein synthesis. (Based on data from van Deurs *et al.*, 1987, 1988 and Endo & Tsurugi, 1987.)

not on the soluble elongation factor and is, therefore, different from that of diphtheria toxin. It was first suggested that the A-chain of ricin may inactivate ribosomes through its ribonuclease activity (Obrig, Moran & Colinas, 1985). However, the ribonuclease activity of the A-chain is effective only with naked ribosomal RNA, but it is inactive with intact ribosomes. In fact, it was shown later that the A-chain enzyme inactivates the ribosomes by cleaving one particular N-glycosidic bond of adenosine[4324] nucleoside, although there are approximately 7000 such bonds present in the 28S rRNA of eukaryotic ribosomes (Endo & Tsurugi, 1987). Similar depurination of the yeast 26S ribosomal RNA was achieved by a recombinant ricin A-chain preparation (Bradley, Silva & McGuire, 1987). A schematic representation of this process is given in Fig. 1.5.

The agglutinin from castor bean, RCA_I, has been studied less extensively. This lectin is a tetramer with a molecular weight value of approximately 120,000. It is composed of two A' and two B' subunits. Ricin and the haemagglutinin have been shown by immunochemical methods (Nicolson *et al.*, 1974) and by amino acid sequencing (Cawley, Hedblom & Houston, 1978; Araki, Yoshioka & Funatsu, 1986) to have similarities in their properties. However, there are also obvious differences. For example, the A-chain of ricin is almost five times more potent as an inhibitor of protein synthesis in cell-free systems than the corresponding A'-chain of RCA_I (Cawley *et al.*, 1978). There is also some sequence homology between the B-chain of ricin and the corresponding B'-chain of the haemagglutinin, despite the differences in 45 amino acid residues between the primary sequences of the two polypeptide chains (Araki *et al.*, 1986). Moreover, there is also appreciable internal homology, about 31%, between the 135 amino acid-containing amino terminal peptide part and the carboxyl terminal polypeptide part, composed of 127 amino acid residues, of the B'-chain. An important difference between the two B-chains is that the B' subunit of RCA_I contains only one sugar-binding site, while the B-chain of ricin has two sites. Consequently, separated B' chains have no cytoagglutinating activity (Nicolson *et al.*, 1974; Shimoda & Funatsu, 1985). Despite all such differences, the structure of the cell-free translation products of poly A-rich RNA of castor bean seeds suggests that a common non-glycosylated polypeptide of about 60,000 molecular weight may serve as a precursor to the A- and B-chains of both ricin and the haemagglutinin (Butterworth & Lord, 1983).

The carbohydrate-binding capacity of the haemagglutinin has been studied extensively and compared with that of ricin. Thus, it is known that lactose is about seven times more effective as an inhibitor of the biological activity of RCA_I than galactose. On the other hand, as it has been mentioned previously, N-acetyl-D-galactosamine is a very poor inhibitor of the haemagglutinin, while the acetylated amino sugar is about 40% as effective as galactose for the inhibition of the biological activity of ricin (Nicolson *et al.*, 1974).

Soyabean (*Glycine max*) agglutinin

This lectin was first isolated by conventional protein purification methods (Wada, Pallansch & Liener, 1958; Lis, Sharon & Katchalski, 1964). It is also one of the first plant glycoproteins described (Dorland *et al.*, 1981), preceded only by the characterization of glycoproteins from kidney bean seeds (Pusztai, 1964). As with most other lectins, purification and isolation is more easily achieved by affinity chromatography (Gordon *et al.*, 1972; Allen & Neuberger, 1975). As the agglutinin from soyabean has affinity for galactose-containing carbohydrate components, guar gum has been suggested as a possible affinity absorbent suitable for its purification (Lonngren, Goldstein & Bywater, 1976). Taking up this suggestion, by incorporating guar gum affinity chromatography, a new and integrated method for the purification of the major anti-nutritional proteins of the seeds of soyabean, which includes the lectin in addition to two different trypsin inhibitors, has recently been worked out (Pusztai, Watt & Stewart, 1991). In this method, the α-galactosides contained in the seed are first removed by extraction with 60% aqueous ethanol. This step is essential as the α-galactosides are natural binders of the soya agglutinin. In their presence the lectins extracted from the seed meal would not bind to the guar gum affinity column. The lectin at pH 5.8 is selectively absorbed on to the cross-linked guar gum affinity column, followed by its specific elution with the same pH 5.8 buffer but also containing 0.1 M galactose (see flow chart, Fig. 1.6). This procedure reduces appreciably the extent of aggregation usually found with soyabean agglutinin purified by other methods (Lotan *et al.*, 1973).

   Soyabean agglutinin preparations are known to contain a number of isolectins, each with a different isoelectric point. Thus, by ion-exchange chromatography, a soyabean agglutinin preparation was separated into four isolectins as early as 1966 (Lis *et al.*, 1966). All components had haemagglutinating activity, were similar in composition and cross-reacted immunochemically. Similar isolectin components, with isoelectric points between pH 5.85 and 6.20, were also obtained by isoelectric focusing (Catsimpoolas & Meyer, 1969). The existence of several isolectins has also been shown more recently by high resolution, flat bed gel isoelectric focusing in the presence or absence of denaturing agents for the soyabean agglutinin preparations isolated by the guar gum affinity method at pH 5.8 (Fig. 1.7). These results confirm the earlier demonstrations of isoelectric heterogeneity and suggest that the isolectins are genuine seed components and not artefacts produced during isolation. Differences in metal ion-bindings have also been eliminated as the source of the charge heterogeneity by showing that the isoelectric banding patterns in denaturing solvents are not altered by extensive dialysis against EDTA dissolved in 8 M solutions of urea (Fig. 1.7).

The existence of isolectins in soyabean agglutinin preparations, however, may also be due to the observed slight size heterogeneity. Indeed,

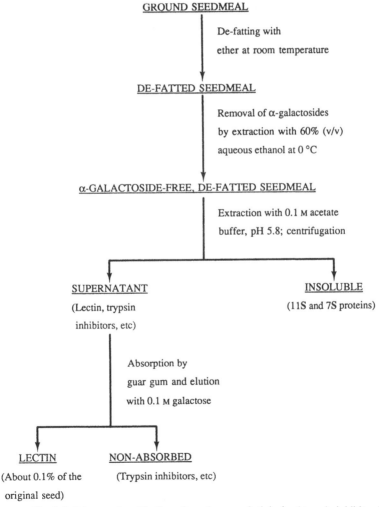

Fig. 1.6. Scheme of purification of soyabean agglutinin (and trypsin inhibitors).

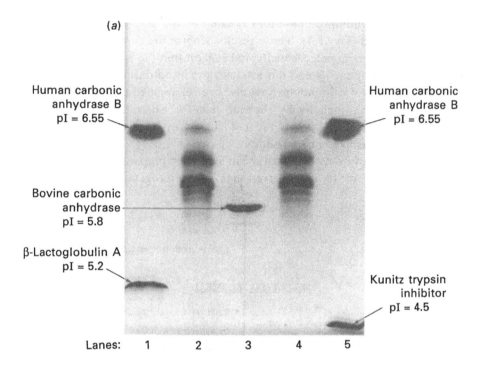

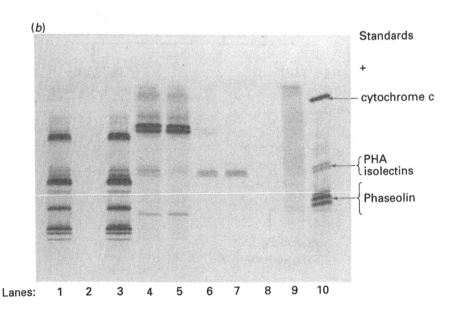

two closely spaced subunit bands have been described in soyabean agglutinin preparations (Lotan *et al.*, 1974, 1975*b*). The presence of two slightly different size subunits has also been confirmed by high-resolution SDS gel electrophoresis for the soya lectin preparations obtained by the guar gum method (Pusztai *et al.*, in press).

The carbohydrate-binding specificity of soyabean agglutinin has been determined from the extent of inhibition of the biological activity of the lectin by carbohydrate components of known structure. Such studies have revealed that the most complementary carbohydrate structures for the sugar-binding site of the agglutinin are some of the glycosides of *N*-acetyl-D-galactosamine or oligosaccharides with a non-reducing terminal *N*-acetyl-D-galactosamine residue (Pereira, Kabat & Sharon, 1974). Galactose, or its derivatives, are less effective inhibitors. It is to be noted, however, that a number of α-galactosidic derivatives of sucrose, such as raffinose, stachyose, verbascose, etc, which are present in soyabean (and other legume) seeds in relatively high concentrations (up 10% w/w) have an inhibitory capacity of about 14% of that of *N*-acetyl-D-galactosamine (Pereira *et al.*, 1974). This explains the need for the removal of these glycosides by extraction with 60% aqueous ethanol from extracts of the soya seed before the soyabean agglutinin can be absorbed onto guar gum (or other) affinity supports (Pusztai *et al.*, in press).

The haemagglutinating activity of soyabean agglutinin can be enhanced by increasing the molecular size of the lectin. Similarly, its mitogenicity against neuraminidase-treated lymphocytes is appreciably enhanced by chemically cross-linking the lectin monomers (Lotan *et al.*, 1973). The same effects may also be achieved by the moderate polymerization obtained on freeze-drying of soyabean agglutinin preparations (Schechter *et al.*, 1976).

Fig. 1.7. Separation of components of soyabean agglutinin preparations by (*a*) isoelectric focusing and (*b*) by SDS-polyacrylamide gel electrophoresis (Pusztai *et al.*, unpublished observations). The following materials were run (*a*) isoelectric focusing: Lane 1, standards, human carbonic anhydrase B (pI = 6.55) and β-lactoglobulin A (pI = 5.2); 2 and 4, purified soya bean agglutinin; 3, standard, bovine carbonic anhydrase (pI = 5.8) and 5, standards, human carbonic anhydrase B (pI = 6.55) and Kunitz trypsin inhibitor (pI = 4.5) and (b) SDS-polyacrylamide gel electrophoresis: Lane 1, soya meal extract; 2, soyafraction soluble in 70% ethanol; 3, soya, insoluble in 70% ethanol; 4, fraction soluble in pH 5.8 buffer; 5, non-absorbed by guar gum; 7, main lectin fraction eluted with 0.1 M galactose from guar gum column; 6 and 8, small amount of materials eluted immediately before and after the main lectin peak, respectively; 9, eluted with 8 M urea from guar gum (aggregated material) and 10, standards, cytochrome c (13 kD), PHA isolectins (30–31 kD) and phaseolin (4 bands of 47, 50, 53 and 54 kD) from the seeds of kidney bean.

In addition to the 'classical' *N*-acetyl-D-galactosamine/D-galactose-specific agglutinin, the seeds of soyabean are now known to contain another lectin. This lectin is unrelated to the more widely studied soyabean agglutinin and interacts specifically with 4-*o*-methyl-D-glucuruno-L-rhamnan from *Rhizobium japonicum* strain 61A76. The lectin can be most conveniently purified from extracts of the soya seeds by affinity chromatography on a support prepared from extracellular poly-saccharide preparations obtained from the bacteria (Rutherford *et al.*, 1986). The agglutinin is a tetrameric glycoprotein with a molecular weight value of about 175,000 and contains two different types of subunit, each of $M_r = 45,000$. The carbohydrate specificity of the lectin has been established and shown to be directed towards 4-*o*-methyl-D-glucuronic acid, D-glucuronic acid and their methyl glycosides. It has been suggested that this lectin is one of the principal components of the recognition process in the *Rhizobium*-legume symbiosis between soyabean roots and the specific nodulating bacteria.

Phaseolus vulgaris lectins (PHA)

Despite earlier claims that haemagglutinating preparations obtained from kidney beans are effectively inhibited by high concentrations of *N*-acetyl-D-galactosamine (Borberg *et al.*, 1966), it is now generally accepted that PHA only recognizes complex carbohydrates of appropriate composition and structure. However, for historical reasons PHA is still usually considered together with other members of this class of lectins.

Lectin preparations from the seeds of kidney beans have first been purified by conventional purification methods (Rigas & Osgood, 1955). As the haemagglutination and lymphocyte-stimulation activities of the various lectin preparations obtained by different workers applying differ-ent protein purification methods showed considerable differences, the two biological activities of the lectin were thought to reside in two different parts of the molecule (Johnson & Rigas, 1972). This has since been confirmed experimentally by the isolation of PHA isolectins in which the two biological activities varied in a predictable manner (Miller *et al.*, 1973; Leavitt, Felsted & Bachur, 1977; Manen & Pusztai, 1982). Thus, it has been shown that in some varieties of *Phaseolus vulgaris* seeds the lectin molecules are made up of two different subunits: E, erythroagg-lutinating and L, lymphoagglutinating subunits. Theoretically, as the lectin is a tetramer, it may occur in the form of five isolectins. Such five isolectins have, in fact, been isolated by ion-exchange chromatography on SP-Sephadex (Leavitt *et al.*, 1977; Manen & Pusztai, 1982).

The molecular weight of the isolectin tetramers is very similar, about 118,000, despite that they can contain two non-identical subunits of different biological activities (Pusztai & Stewart, 1978). Indeed, the molecular weight of both the E and L subunits separated by isoelectric focusing in dissociating solvents is very close, about 30,000 (Miller *et al.*, 1973, 1975; Pusztai & Stewart, 1978).

In early studies, most of the differences in the primary sequence of the two different subunits have been found in the first seven amino acid residues of the amino terminal end of the polypeptide chains. The rest of the sequence from there on, up to the 24th amino acid residue, is the same for both the E and L subunits (Miller *et al.*, 1975). More recently, the entire primary amino acid sequence of PHA has been determined from the nucleotide sequences of cDNA obtained from lectin mRNA isolated from the seeds and by the use of reverse transcriptase (Hoffman, Ma & Barker, 1982). It is clear that, although the two different subunits, E and L, are coded by two different genes, they are closely linked on the chromosome (Hoffman & Donaldson, 1985). Moreover, extensive homologies between the primary structure of the two subunits (and that of other legume lectins) have also been found (Strosberg *et al.*, 1986*a*, *b*).

Rabbits parenterally immunized with pure E_4 and L_4 PHA isolectins give rise to antibody preparations which cross-react extensively. It is, however, possible to prepare monospecific pure IgG antibodies for the E_4 or L_4 isolectins from these cross-reacting antibody preparations by absorption of the unwanted contaminating antibodies with L_4-Sepharose-4B and E_4-Sepharose-4B affinity absorbents, respectively (Manen & Pusztai, 1982; Felsted *et al.*, 1982). Monospecific anti-E or anti-L antibody preparations thus obtained have been used for the localization of both E- and L-type lectins in the cotyledonary or axis cells of the *Phaseolus vulgaris* seeds. The results have shown that in cells of the seeds of kidney bean, the lectins, both E and L types, are confined to protein bodies. However, in axis cells, lectins are also found outside these subcellular organelles (Manen & Pusztai, 1982).

It has been suspected for some time that the above simple model of five isolectins, based on the existence of two different, E and L, subunits is not valid generally for all cultivars of kidney bean. There are over 30,000 varieties of kidney beans known and deposited in seed banks. Most of these can contain a multiplicity of PHA subunits of various types. It is also known that some varieties may be devoid of the L subunit entirely. These, unless tested under special conditions, show very little mitogenic activity (Pusztai & Watt, 1974; Pusztai *et al.*, 1983*a*).

About 7–10% of kidney bean varieties were thought not to contain lectins. However, this is not strictly correct as from one such kidney bean variety, Pinto III, a small amount of a lectin of low haemagglutination activity against rabbit erythrocytes has now been isolated (Pusztai, Grant & Stewart, 1981a). The Pinto lectin is related to the generally occurring PHA-types found in most seeds, although it is a dimer and not a tetramer, like most common PHA isolectins. Furthermore, the carbohydrate-binding specificity of the Pinto lectin is different from that of the usual PHA isolectins as, for example, it does not bind to fetuin or thyroglobulin at all (Pusztai, Grant & Stewart, 1982). However, despite all these differences, the three main types of PHA isolectins (Fig. 1.8) from the various cultivars of kidney beans, whether they contain both E and L subunits or only E subunits, are shown by their immunochemical cross-reactivity (Pusztai et al., 1983a) to be related. This is even true for the Pinto III-type PHA isolectins.

Jaffe originally classified kidney beans on the basis of their haemagglutination activity with erythrocytes from different animal species into four groups: A, B, C or D (Jaffe, Levy & Gonzales, 1974). Thus, group A lectins are very active with cells obtained from most animal species. Group B PHA's react with most cells, except trypsin-treated cow erythrocytes. On the other hand, group C kidney bean lectins are inactive against human or rabbit red cells but agglutinate cow cells strongly. Finally, group D lectins show very little activity with practically all untreated cells. Their haemagglutination activity is manifested only when pronase-treated hamster or mouse erythrocytes are used (Jaffé et al., 1974).

More recently, these observations of differences in haemagglutination activity of PHA lectins from different cultivars have been put on a solid protein chemical foundation by Brown and his colleagues. Not surprisingly, it has been found that lectins in the four groups originally proposed by Jaffé have different and characteristic subunit patterns (Brown et al., 1982). By using two-dimensional separation methods based on isoelectric focusing in dissociating solvent, followed by SDS-polyacrylamide gel electrophoresis, over one hundred varieties of *Phaseolus vulgaris* seeds have been examined. Rather conveniently, some of the seeds examined also included a number of cultivars on which the original classification by Jaffé was based. The relationship between the distinct polypeptide patterns obtained by the two-dimensional electrophoretic analyses and the agglutination ratios for cow *vs* rabbit erythrocytes suggested that kidney bean lectins may indeed be classified into four distinct groups (Brown et al., 1982). However, according to their findings, the subunit patterns of even the group A lectins, which are most commonly found in kidney bean cultivars and come closest to the simple scheme of five isolectins based on E and L subunits, are a great deal more complex than originally proposed (Osborn *et al.*, 1983). Thus the L subunit in reality contains two major, but different, polypeptides. Similarly, the E subunit, instead of a single polypeptide, is made up of one major and two minor polypeptides. Clearly, the various combinations of all possible different polypeptides may give rise to a far greater number of isolectins than the five originally proposed (Osborn *et al.*, 1983).

The reasons for the diversity and complexity of the lectin polypeptide subunits occurring in the multitude of *Phaseolus vulgaris* cultivars are not

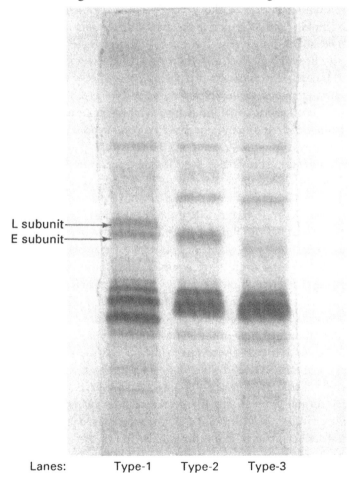

Fig. 1.8. SDS-polyacrylamide gel electrophoresis patterns of the three basic types of lectins from the seeds of *Phaseolus vulgaris*. Type-1 PHA contains both E and L subunits, while type-2 PHA is based on the E subunit. Type-3 beans contain neither E nor L subunits but very small amounts of a different lectin, designated as Pinto III-type PHA (not clearly visible because of the low lectin content).

known. Obviously, some variations in gene structure coding for PHA lectins may have occurred during evolution. However, modifications in the genes are permissible only within an overall general constraint, i.e. the structural features responsible for the biological and plant physiological functions of the lectins must remain intact. Moreover, cultivation by man has meant selection for desired characteristics. This has been achieved by extensive crossing and breeding of cultivars in which the putative, simple ancestral polypeptide(s) have given place to much more complex subunit polypeptide patterns. In addition, the *Phaseolus* genus contains a number of species closely related to *Phaseolus vulgaris*, such as, for example, *Phaseolus coccineus* and/or *Phaseolus acutifolius*. As these can be successfully crossed with most *Phaseolus vulgaris* cultivars, the genetic diversity for lectins within the *Phaseolus* genus is further increased. Indeed, lectins, similar to those in *Phaseolus vulgaris* have also been isolated from *Phaseolus coccineus* seeds (Novakova & Kocourek, 1974). Also, both erythroagglutinins and lymphoagglutinins have been isolated from the seeds of tepary beans (*Phaseolus acutifolius*) by affinity chromatography on fetuin-Sepharose-4B (Pusztai, Watt & Stewart, 1987) by a procedure previously used for the preparation of *Phaseolus vulgaris* isolectins (Pusztai & Palmer, 1977; Pusztai & Stewart, 1978). The tepary bean isolectins are also tetramers, but their molecular weight, about 116,000, is somewhat less than that of the corresponding kidney bean isolectins. The subunit polypeptide patterns of the *Phaseolus acutifolius* isolectins established by two-dimensional peptide mapping has been shown to be even more complex than those of kidney bean isolectins (Fig. 1.9). No simple scheme, based on E- and L-type subunits, can account for their great complexity (Pusztai *et al.*, 1987).

The biological activities of isolectins from *Phaseolus vulgaris* seeds and those of the other members of this genus cannot be effectively inhibited by simple sugars but only by complex oligosaccharide structures. The great biological effectiveness of the kidney bean isolectins is the result of their specific recognition of and binding to oligosaccharide structures which are normal components of the membranes of most eukaryotic cells. Thus, studies of Kornfeld and his associates on glycopeptides obtained from erythrocyte membranes have shown the involvement of β-galactosyl groups in the specificity of PHA (Kornfeld, Gregory & Kornfeld, 1972). Similar results have also been obtained with glycopeptides obtained from thyroglobulin (Irimura *et al.*, 1975). By using quantitative precipitation–inhibition methods, Hammarström *et al.* (1982) have found that the most complementary oligosaccharide to inhibit the activity of the L subunit-containing PHA is a pentasaccharide, Galβ1,4-

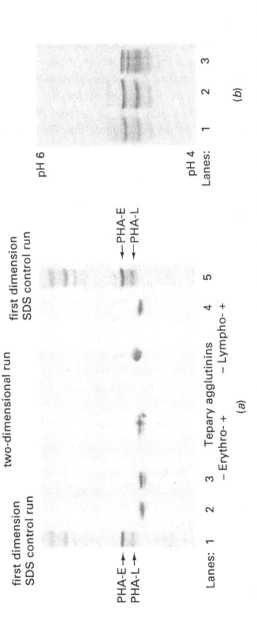

first dimension
SDS control run

two-dimensional run

first dimension
SDS control run

PHA-E →
PHA-L →

← PHA-E
← PHA-L

Lanes: 1   2   3   4   5
        – Erythro– +   – Lympho– +
              Tepary agglutinins

(a)

pH 6

pH 4

Lanes: 1   2   3

(b)

Fig. 1.9. Patterns of separation of the erythro- and lymphoagglutinins of *Phaseolus acutifolius* by (*a*) two-dimensional electrophoresis and (*b*) isoelectric focusing. The erythroagglutinin and lymphoagglutinin samples from tepary bean were first separated by isoelectric focusing in the pH range of 4–6. The lanes corresponding to these samples were then cut out from the gel and run in SDS-polyacrylamide gel electrophoresis at right angles to the direction of separation in the first dimension. To aid the correct assignment of the bands, several standards were included in both runs: (*a*) SDS-polyacrylamide gel electrophoresis; Lanes 1 and 5, kidney bean seed extract; 2 and 4, erythroagglutinin and 3, lymphoagglutinin from tepary bean and (*b*) isoelectric focusing; Lanes 1 and 2, two samples of lymphoagglutinin and 3, erythroagglutinin from tepary bean (Pusztai *et al.*, 1987; by permission of *Phytochemistry*).

GlcNAcβ1,2-(Galβ1,4GlcNAcβ1,6)-Man, found in tetrabranched and some tribranched oligosaccharides. Similar conclusions have been drawn from studies with glycopeptides on L-PHA affinity columns (Cummings & Kornfeld, 1982). Accordingly, there is a requirement for the presence of a galactose residue in the glycopeptides before they are significantly retarded by affinity columns containing PHA lectins made up of either the E or the L subunits. A bisecting *N*-acetyl-D-glucosamine in the oligosaccharide has been found to be important for the binding of complex carbohydrates to E-PHA but not to L-type PHA-affinity columns. Indeed, the conclusion seems inescapable that the specificities of the E- and L-type PHA lectins must be grossly similar (Hammarström *et al.*, 1982). Similar conclusions have been reached from the results of experiments with affinity columns containing immobilized PHA lectins (Green & Baenziger, 1987) in which the free and reduced oligosaccharides released from the glycoprotein–asparagine linkage after treatment with *N*-glycanase and reduction with sodium borohydride, have been used as ligands. Thus, the E and L lectins have virtually identical specificities for the reduced oligosaccharides. Both lectins can retard α-2,3- but not α-2,6-linked sialic acid-containing saccharide components. De-sialylated oligosaccharides containing one to four peripheral *N*-acetyllactosamine-type branches are also retarded by the columns. Removal of the galactose residue substantially reduces or totally eliminates the binding of the oligosaccharide. A bisecting *N*-acetyl-D-glucosamine residue attached to the β-linked core mannose shows very high activity with both E and L-type PHA's. Thus, these studies have confirmed the essential nature of the branch β-galactosyl groups in non-reducing terminal position for the biological activity of kidney bean lectins.

L-fucose-binding lectins

This is a relatively small group of lectins. Some of them are useful reagents for the detection of fucose residues in complex carbohydrate structures. Moreover, some of these lectins, such as the *Ulex europaeus* I lectin, are used as blood group O-specific reagents. For example, an anti-H(O) lectin has been purified from gorse seeds. These seeds, however, contain another lectin, called *Ulex europaeus* II lectin, which is inhibited by a number of β-glycosides, such as cellobiose, lactose or chitin oligosaccharides and is, therefore, completely different from the fucose-specific lectin from the same seeds (Pereira, Gruezo & Kabat, 1979).

The fucose-specific *Ulex europaeus* I lectin has been purified by affinity chromatography. Several affinity supports have been found useful for the isolation of the lectin. These include *o*-α-L-fucopyranosyl polyacrylamide gel (Horejsi & Kocourek, 1974) or 6-aminohexyl α,β-λ-fucopyranoside coupled to Sepharose-4B gel (Frost *et al.*, 1975). The lectin is a dimer glycoprotein with a molecular weight value between 60,000 and 68,000. It contains two dissimilar subunits with M_r = 29,000 and 31,000, respectively (Frost *et al.*, 1975). In contrast, other workers have reported the presence of only one subunit of 40,000 to 42,000 molecular weight in SDS solutions (Horejsi & Kocourek, 1974; Pereira *et al.*, 1978). The lectin contains several metal ions (Horejsi & Kocourek, 1974), which are apparently not necessary for haemagglutination activity.

The lectin shows a strong preference for 2-*o*-linked α-L-fucosyl groups. Thus, 2′-*o*-α-L-fucosyllactose is a good inhibitor, while the corresponding 3′ or 6′-*o*-α- or β-linked derivatives are not. The best inhibitor of the *Ulex europaeus* I lectin is the type-2 H active oligosaccharide structures of L-fucosyl α-(1–2)galactosylβ-(1–4)*N*-acetylglucosaminoylβ-(1–6) derivatives obtained from blood group substances of H(O) specificity. The type-1 H active oligosaccharides, on the other hand, are poor inhibitors (Pereira *et al.*, 1978). The strict specificity of the gorse lectin is also shown by the fact that it has a high affinity for L-fucose, while L-arabinose or most hexoses do not inhibit its haemagglutination activity (Frost *et al.*, 1975). The lectin precipitates blood group H-specific substances quite strongly. Its reactivity with A_2-specific substances is less strong. Moreover, the gorse lectin is virtually inactive with A_1, B or Le^a substances. The poor reactivity with blood group Le^a-specific sugar structures shows up the differences in the specificities of the two major fucose-reactive lectins. Thus, while the gorse lectin is inactive, the other fucose-specific lectin from *Lotus tetragonolobus* is a good reagent for Le^a-specific red cells.

2

Structure of lectins

Most of our knowledge of the primary structure of plant lectins has been gathered in the last decade from extensive chemical studies with lectins isolated from leguminous plants. The comparative, evolutionary and chemotaxonomical aspects of such studies are of particularly special value for our understanding of the potential functional significance of lectins and gene alterations in plants during evolution.

Primary sequences

A comparison of partial or complete primary sequences of lectins of diverse origin, including one- or two-chain lectins, has shown up extensive homologies between different lectins (Strosberg *et al.*, 1986*a*,*b*). Thus, for example, the amino terminal sequences of the β-chains of two-chain lectins from Vicieae and of one-chain lectins from other tribes of Papilionaceae or Caesalpinoideae, compared with the appropriate sequence of concanavalin A (residues of 123–147), are highly homologous (Strosberg *et al.*, 1983; 1986*a*,*b*). It has also become clear from sequence comparisons that lectins from the same tribe are more homologous to each other than to members of other tribes. Rather interestingly, phenylalanine residues in positions 6 and 11 are conserved in all 26 lectins used for comparison studies. Residue 5 is also found to be highly conserved and occurs in 23 of the 26 lectins studied.

Thus, the comparison of even partial sequences has turned out to be very rewarding. However, more is to be expected from comparisons of the structure of fully sequenced lectins, and the value of such structural comparisons is likely to grow in future. Up until now, Strosberg and his colleagues have already compared the full primary amino acid sequences of seven leguminous lectins. Four of these, including concanavalin A

(Edelman *et al.*, 1972) and lectins from *Lens esculenta* (Foriers *et al.*, 1981), sainfoin, *Onobrychis viciifolia* (Kouchalakos *et al.*, 1984) and broad bean, *Vicia faba* (Cunningham *et al.*, 1979) have been sequenced by classical methods. The sequences of the other three lectins, such as PHA from kidney bean (Hoffman *et al.*, 1982), soyabean agglutinin (Vodkin, Rhodes & Goldberg, 1983) and pea, *Pisum sativum*, lectin (Higgins *et al.*, 1983*a*) have been deduced from cDNA sequences. The results of the comparison of the full sequences of these lectins have fully supported the conclusion reached previously from partial sequence comparisons. Accordingly, when appropriate deletions and gaps are taken into account, and providing that the primary amino acid sequence of concanavalin A is circularly permuted, i.e. rearranged into two portions, all seven lectins of Leguminoseae are homologous in their entire primary sequences (Strosberg *et al.*, 1986*b*). For reasons detailed previously (Chapter 1, p. 5), the required rearrangement of the sequence of concanavalin A to maximize the best alignment with the other six lectins has a solid biological foundation. Concanavalin A is synthesized as a single glycoprotein precursor which is then subject to post-translational processing (Carrington *et al.*, 1985). During this, there is a midchain scission of the polypeptide chain and the resulting two fragments are religated in a different order (Fig. 1.1). Thus, it is not surprising that maximum homology to the two-chain lectins from Vicieae is obtained when their β-chains are aligned with residue no. 120 and their α-chains run from residues 70 to 119 of those in concanavalin A (Olsen, 1983; van Driessche, 1987; Fig. 2.1).

Comparisons of primary sequences have recently been extended to include lectin sequences of α-chain subunits from 13 additional two-chain lectins, or isolectins from the tribe of Vicieae with homologous regions of single-chain lectins from other tribes (Rouge *et al.*, 1987). The results have provided convincing evidence that differences are few and confined to few amino acids in the α-chains of 16 two-chain lectins and isolectins of Vicieae, for which sequences are known. However, when these comparisons are made with the appropriate regions of single-chain lectins from other tribes, the number of differences is increased, despite the extensive homologies. It is particularly instructive to set out the percentages of sequence identities in the lectins and the minimum number of nucleotide substitutions required to interconvert the differences found to identical sequences (Table 2.1). The results thus demonstrate the striking homologies which exist within the primary sequences of Vicieae lectins and they also provide a quantitative estimate of the extent of differences with other lectins.

Table 2.1. *Comparison of percentage identities in homologous legume lectins.*

		a	b	c	d	e	f	g	h	i	j	k	l	m	n	o	p	q	r
a	*Lathyrus articulatus*	0	96	91	86	87	88	87	87	85	81	81	77	79	58	40	43	42	45
b	*Lathyrus ochrus α1*	2	0	92	84	85	86	85	83	83	79	79	77	75	55	36	42	40	43
c	*Lathyrus ochrus α2*	5	4	0	82	83	82	81	83	85	81	81	77	77	55	36	42	42	43
d	*Lathyrus cicera α1*	7	11	12	0	97	93	95	87	92	83	83	76	78	56	41	44	45	45
e	*Lathyrus cicera α2*	8	9	10	1	0	95	92	85	91	85	85	77	79	55	43	45	46	46
f	*Lathyrus odoratus*	10	11	13	4	5	0	92	88	91	86	86	78	80	54	43	44	45	45
g	*Lathyrus sativus*	12	13	16	5	9	7	0	85	88	83	83	77	77	53	40	44	42	47
h	*Lathyrus aphaca*	7	9	10	8	9	10	14	0	84	81	79	75	73	53	45	47	40	45
i	*Lens culinaris*	13	14	13	12	9	12	16	14	0	78	78	76	76	55	42	46	43	44
j	*Pisum sativum*	12	13	14	13	9	11	16	12	16	0	96	79	83	60	45	47	44	45
k	*Vicia cracca*	12	13	14	10	9	11	16	13	16	2	0	79	83	60	43	47	42	43
l	*Vicia faba*	19	20	22	22	19	21	24	20	18	17	16	0	75	56	45	49	42	45
m	*Vicia sativa*	18	20	22	21	18	20	24	20	19	16	16	15	0	59	43	47	46	45
n	*Glycine max*	31	29	29	36	35	36	37	33	33	33	34	33	32	0				
o	*Canavalia ensiformis*	45	47	47	46	43	44	47	42	43	39	40	39	40		0			
p	*Dioclea grandiflora*	41	43	42	45	43	45	46	39	39	40	41	37	37			0		
q	*Onobrychis viciifolia*	45	43	44	45	45	48	48	48	41	43	43	44	45	44	45		0	
r	*Phaseolus vulgaris*	54	53	53	54	53	56	58	51	52	51	51	47	53					0

Note: Calculated percentage identities (above diagonal) and minimum number of nucleotide substitutions (under diagonal) required to interconvert the differences to identical sequences. 0 = complete identity.

Source: From Rouge *et al.*, 1987; by permission of authors and publisher.

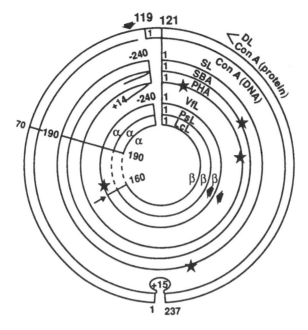

Fig. 2.1. Schematic representation of two-chain lectins, one-chain lectin and concanavalin A showing circular permutation which gives maximal homology (from van Driessche, 1987; reproduced by permission of the author). The heavy arrows represent natural cleavage sites in what would be a continuous peptide sequence of 240 amino acids, such as it is in SL, SBA or PHA. In VfL, PsL and LcL the two subunits, α and β, are the scission products as indicated schematically by the arrows. Similarly, the heavy asterisks represent the putative glycosylation sites on the 240-long peptide sequence. Those which have no such asterisk are not glycoproteins. The broken lines indicate stretches of missing amino acids in the sequence. Abbreviations: DL, *Dioclea grandiflora* lectin; *con A*, concanavalin A, sequence derived from protein sequencing (protein) or DNA sequencing (DNA); SL, sainfoin lectin (*Onobrychis viciifolia*); SBA, soyabean agglutinin; PHA, kidney bean lectin; VfL, *Vicia faba* lectin; PsL, *Pisum sativum* lectin; LcL, *Lens culinaris* lectin.

Metal binding sites

Most leguminous lectins studied contain metal ions as an essential part of their overall native structure. Metal ions have also been shown to be important for biological activity. In concanavalin A, the metal binding sites are situated in the amino terminal part of the polypeptide chain. Rather interestingly, the amino acids in appropriate sequences involved in the binding of metal ions have been highly conserved in six out of the seven lectins compared. In concanavalin A the amino acid residues involved in the binding of one calcium and one manganese ion per subunit are well known. These include two aspartic acid residues (asp 10 and 19), one residue of each, asparagine (asn 14), histidine (his 24), serine (ser 34) glutamic acid (glu 8) and tyrosine (tyr 12). In comparable sequences of lectins from soyabean, peas, faba bean, lentils and sainfoin, but not in PHA from kidney bean, all these amino acids involved in metal binding have been conserved with the exception of the tyrosine residue at position no. 12 of concanavalin A, which is replaced by phenylalanine in the other legume lectins. Furthermore, in the lectin from *Dioclea grandiflora* which, like *Canavalia ensiformis*, is a member of the Diocleae tribe, all residues of the metal binding site of concanavalin A have been strictly conserved, with the exception of the aspartic acid residue at position 10, which is substituted by asparagine (Richardson *et al.*, 1984). In the phylogenetically more distant PHA only three residues of the metal binding site of concanavalin A have been conserved. These include glutamic acid (glu 8), aspartic acid (asp 10) and serine (ser 34), while the histidine residue at position 24 is replaced with arginine.

Hydrophobic sites

Hydrophobic interactions have been shown to play a very important role in the maintenance of the native structure of most lectins studied so far. Hydrophobic sites also make an important contribution to ligand binding, particularly ligands of hydrophobic character. In concanavalin A there is a well-defined hydrophobic cavity which is made up of a number of hydrophobic amino acid residues in the primary polypeptide sequence. The side chains of leucine (leu 81), valine (val 89) and three phenylalanine residues (phe 111, 191 and 212) constitute the main part of the hydrophobic cavity. As all these amino acid residues are strictly conserved in most lectins studied up to date (Strosberg *et al.*, 1986*a*), the essential nature of hydrophobic interactions for the conformation and ligand binding of the lectins is clearly shown. When the amino acids, which contribute to the hydrophobic cavity in concanavalin A, are not conserved in other lectins, they are always replaced with other hydrophobic amino acids.

The existence of such hydrophobic cavities in the structure of lectins may have important biological consequences. Thus, for example, the binding through the hydrophobic site of auxins (Edelman & Wang, 1978) or cytokinin and adenine (Roberts & Goldstein, 1983*a*,*b*) by concanavalin A may have important bearings on the possible functions of this lectin in the life cycle of the plant. As previously mentioned, concanavalin A binds *myo*-inositol quite independently of its specific interaction with D-mannose/D-glucose (Wasseff *et al.*, 1985). It is also possible that the strong binding of concanavalin A to liposomes via phosphatidylinositol may be considerably strengthened by hydrophobic interactions between the apolar amino acid side-chains of the lectin and lipid components of the liposomes.

Glycosylation sites

Although most lectins are glycoproteins, there are a number of well-known lectins, such as, for example, concanavalin A, lentil lectin or wheatgerm agglutinin, which contain no covalently attached carbohydrates. Rather interestingly, however, even the non-glycoprotein lectins are usually synthesized as glycosylated precursors. Thus, proconcanavalin A is an inactive glycoprotein from which the glycosidic side-chain is removed during post-translational processing (Bowles *et al.*, 1986). Similarly, the non-glycoprotein wheatgerm agglutinin molecule is produced by removing a carboxyl terminal glycopeptide from the glycosylated precursor during post-translational processing (Mansfield, Peumans & Raikhel, 1988).

All glycoprotein lectins contain a peptide sequence, asparagine-X-threonine/serine, which is characteristic for glycosylation sites. These sequences are different in the non-glycoprotein lectins. Thus, in the β-chain of *Vicia faba* lectin (favin) the sequence to which the glycosyl side-chain is attached contains the peptide asparaginyl(169)–alanyl–threonine(171). In the non-glycoprotein, concanavalin A, this is replaced by asparaginyl–serinyl–valine. Peptide sequences, which in one glycoprotein lectin contain the glycosidic side-chains, are not necessarily conserved in another glycoprotein lectin. Accordingly, in addition to other experimental evidence not related to these studies, these results also suggest that the carbohydrate part of the lectin structure may not have much functional significance for the biological activity of the lectins.

Carbohydrate-binding sites

As lectins, even those belonging to the same carbohydrate-binding class, differ appreciably in the fine specificity of their sugar-binding, it is hardly surprising that the amino acid sequences participating in the carbohydrate-binding site of, for example, concanavalin A (Becker *et al.*, 1976) are poorly conserved in other lectins.

Three-dimensional structure

The structural similarities and homologies found in legume lectins are also expressed in terms of similarities in their three-dimensional structures. For example, a comparison of the predicted secondary structures of the Vicieae lectins calculated by using computerized methods (Chou & Fasman, 1978; Garnier, Osguthorpe & Robson, 1978) with those of concanavalin A, showed the presence of identical or homologous β-turn structures in all these lectins (Rouge *et al.*, 1987). Similarly, the hydropathic profiles of two-chain or single-chain lectins calculated by a computer program (Kyte & Doolittle, 1982) can be easily superimposed (Rouge *et al.*, 1987). The three domains in the structure of concanavalin A as proposed by Olsen (1983) also appear to exist in members of the two-chain lectins of Vicieae. The good fit between the structures of these lectins, however, applies to the already discussed circular permutation of their primary sequences (Fig. 2.1).

Conclusions
The high degree of homology in the primary sequences and three-dimensional structures of both single-chain and two-chain legume lectins suggests that genes coding for them have been modified relatively slightly during evolution. This applies particularly to the highly homologous Vicieae lectins. For differences in these, the simplest type of mutations, i.e. point mutations, give an adequate explanation (Rouge *et al.*, 1987). Because of this close evolutionary relationship, lectins can be used as good phylogenetic markers to study the molecular biology of genes during the speciation process (Rouge *et al.*, 1987).

3

Localization and biosynthesis in plants

Although plant lectins have been known for the last 100 years, we are only beginning to understand their functions in the life-cycle of plants. The fact that lectin genes have been conserved well during evolution, especially in Leguminoseae, and that lectin proteins coded for by the genes are homologous, argues very strongly that lectins may have important, albeit mainly unknown, roles in the life of the plant. In the past, many such roles have been proposed. Thus lectins may act as defensive agents, carbohydrate transporters, recognition agents, storage proteins or growth regulators, etc (for references see Pusztai *et al.*, 1983*a*; Etzler, 1986). However, experimental evidence in support of such roles is still scanty at best.

Whatever role(s) lectin may play, it is important to find out their localization and distribution in various parts of the plant. Temporal changes in their distribution may also help to formulate ideas about the functions of lectins. In addition, any proposed role will also have to be compatible with their ultrastructural localization and any possible changes in it during the various stages of plant development. Owing to their great importance, studies on the localization of lectins have been reviewed extensively (Pusztai *et al.*, 1983*a*; Etzler, 1985, 1986).

Localization in the plant and cellular location

Seeds

Lectins are generally most abundant in the seeds of plants, especially in the seeds of Leguminoseae and Graminaceae. However, lectins are by no means confined to seeds. Variable, although usually small amounts of lectins or lectin-like, immunochemically cross-reacting proteins (CRM) are also present in vegetative parts of plants.

Seed lectins in leguminous plants are located mainly in the cotyledons. Smaller amounts are occasionally found in embryos and even in seed coats (Howard, Sage & Horton, 1972; Pueppke *et al.*, 1978). In some plants, lectins are absent in immature seeds. However, most plants begin to synthesize lectins some time after anthesis. From then on, lectins accumulate parallel with the storage proteins of the seeds (Pusztai *et al.*, 1983*a*). Within the cells of the seeds, lectins are located mostly, if not exclusively, in protein bodies. Thus, by ultrastructural localization methods, the presence of lectins in protein bodies has been confirmed for the seeds of *Phaseolus vulgaris* (Manen & Pusztai, 1982); *Canavalia ensiformis* (Herman & Shannon, 1984*a*), *Glycine max* (Horisberger & Vonlanthen, 1980) and *Pisum sativum* (van Driessche *et al.*, 1981). However, in the vascular cells and embryos of *Phaseolus vulgaris*, both E and L isolectins are also found outside the cytoplasmic protein bodies (Manen & Pusztai, 1982). Similarly, studies based on subcellular fractionation methods, both in aqueous (Pusztai *et al.*, 1977; Pusztai, Stewart & Watt, 1978) and non-aqueous media (Bollini & Chrispeels, 1978; Begbie, 1979), have confirmed the association of E and L isolectins with the protein bodies of the cotyledonary cells of kidney bean seeds. Pulse-chase and other studies have given similar results (Chrispeels, 1984). Subcellular localization studies with the seeds of *Dolichos biflorus* by immunofluorescence, immunocytochemistry and subcellular fractionation methods have also indicated that most of the seed lectin is found in protein bodies. However, the lectin appears to have an uneven distribution within the protein bodies and is concentrated at the periphery of these subcellular organelles. In addition, some of the lectin is associated with starch granules. However, the significance of this is not clear (Etzler *et al.*, 1984).

Lectins are known to be synthesized on the rough endoplasmic reticulum (Bollini & Chrispeels, 1979) and translocated into the lumen (Chrispeels *et al.*, 1982). Although it is not entirely clear yet how the lectin is transmitted to the protein bodies for deposition, it appears that, before reaching the protein bodies, the L-PHA isolectin is first transported through the Golgi stacks, where it becomes fucosylated (Chrispeels, 1983*a*,*b*). Similar conclusions have been reached for the involvement of the Golgi apparatus in the deposition of the seed lectin of *Bauhinia purpurea* (Herman & Shannon, 1984*b*). In fact, the orientation of the Golgi apparatus in these seeds is such that the secretion vesicles of this organ are in the close proximity of the protein bodies (Herman & Shannon, 1984*b*). Indeed, similar considerations apply also to the intracellular transport and deposition of the α-galactosidase–haemagglutinin in developing soyabean cotyledons (Herman & Shannon, 1985). Although this enzyme is not usually considered as a true lectin, as it is immunologically related to galactose-binding legume lectins (Hankins, Kindinger & Shannon, 1980*a*,*b*), its intracellular transport may still be relevant. The Golgi apparatus, in addition to processing the bulk of the storage proteins–lectins, is also responsible for the synthesis and assembly of the thickening walls of parenchymal cells which occur at the same time. Obviously, as the final destination of the cell wall precursors, storage proteins and lectins is different, some form of product-sorting mechanism must exist in the Golgi. Experimental proof for the existence of such a process has been obtained from experiments in which the intracellular traffic of products of the Golgi is disrupted by the addition of the ionophore, monensin. Under these conditions, probably as a result of being packaged into the 'wrong' vesicles, lectin and storage protein precursors are misrouted to the cell surface (Bowles & Pappin, 1988). It, therefore, appears that the involvement of the Golgi apparatus is obligatory in modifying and directing the synthesized lectin to protein bodies, as its ultimate location in cotyledonary cells.

In other dicotyledonous plants, such as, for example, *Datura stramonium*, the results of early work has indicated an essentially similar localization of the lectin (Kilpatrick, Yeoman & Gould, 1979). Thus, examined by differential centrifugation methods, over half of the seed lectin is found associated with protein bodies in the disrupted cells of *Datura stramonium*. However, specific lectin activities of other subcellular fractions are also similar, suggesting that a part of the seed lectin may be entrapped in membrane vesicles (Kilpatrick *et al.*, 1979). Indeed, recent work by immunocytochemical methods done in conjunction with agglutination techniques, has shown that the seed lectin in *Datura stramonium* seeds occurs predominantly in outer seed tissues of the seed coat and epidermis. In these structures the lectin appears to be associated, at least in part, with cell walls (Broekaert *et al.*, 1988). Seeds secrete lectin polypeptides into the surroundings both on development and imbibition. It has been suggested repeatedly that this surface-located lectin may be implicated in the defence of the plant against microbial attack (Sequeira, 1978; Broekaert *et al.*, 1988). Rather interestingly, the potato tuber lectin, which is structurally closely related to the lectin from *Datura stramonium*, is also associated with cell walls and is easily solubilized from this location (Casalongue & Pont Lezica, 1985). A development-dependent lectin in cucumber seeds, specific for *N*-acetyl-D-glucosamine oligomers, has also been found located on epicuticular surfaces (Skubatz & Kessler, 1984). By immersing the seeds in water, this lectin is released quickly (within one min). Although for the first two days of germination there appears to be no lectin in the seeds, its content on the surface rapidly rises after this period and reaches a maximum on about the fourth day. As the presence of this lectin can even be shown in 17 days old seedlings, it may be involved in host–microbial pathogen recognition and defence of the cucumber plant (Skubatz & Kessler, 1984).

The location of seed lectins has also been studied extensively in monocotyledonous plants and, of these, one of the most studied is the wheat plant (*Triticum vulgare*). Wheatgerm agglutinin is present only in small quantities in dry wheat grains where it is confined to the embryo, while the endosperm of the seed contains none (Mishkind, Keegstra & Palevitz, 1980; Miller & Bowles, 1982). Moreover, the lectin is synthesized *in vitro* only by the embryo and not by the endosperm (Triplett & Quatrano, 1982). These results have since been confirmed and refined by the use of immunocytochemical methods (Mishkind *et al.*, 1982). Identical lectins have been found in all members of the Triticeae tribe and in two genera from other tribes of this family of monocotyledonous plants, *Brachypodium* and *Oryza* (rice). They are all dimers, composed of two identical subunits and have the same specificity for N-acetyl-D-glucosamine oligomers (Mishkind *et al.*, 1983). However, despite the close identity of these lectins, their location in wheat, rye or barley appears to be different. For example, there is no lectin present in the coleoptiles of barley, while, in wheat, the lectin is found on the outer surface of this organ. In rye, on the other hand, the lectin is located on both inner and outer surface cells of the coleoptiles. In contrast, the somewhat different rice lectin is distributed throughout the coleoptiles (Mishkind *et al.*, 1983).

The time sequence of the appearance of wheatgerm agglutinin in the tissues during embryogenesis has been established by immunocytochemistry. Thus 8–10 days after anthesis, the lectin begins to accumulate in the radicle, root cap and coleorhiza in the early stage II, prior to the main period of embryo growth. During maximum growth of the embryo, stage III, the agglutinin is first found in the epiblast, followed by its appearance in the coleoptile at a later stage. After this, in stage IV of the development, desiccation occurs (Raikhel & Quatrano, 1986).

There is some experimental evidence to indicate that the accumulation of wheatgerm agglutinin is modulated by the plant hormone, abscisic acid. Similar observations have been made with barley development. Thus, when mature, lectin-free embryos of barley are treated with this hormone, a wheatgerm agglutinin-like protein begins to accumulate in barley coleoptiles (Raikhel & Quatrano, 1986). Indeed, it appears that both the synthesis of the lectin (Raikhel, Palevitz & Haigler, 1986) and the expression of wheatgerm lectin genes (Raikhel & Wilkins, 1987) are under the control of abscisic acid in both embryos and adult wheat plants.

As the lectins in Triticeae are all concentrated at the external surfaces of the embryo, it has been suggested that such a location is compatible with the proposed role for the lectin as a defensive agent against fungal pathogens (Mishkind *et al.*, 1982). It is quite striking that most lectins with specificities for *N*-acetyl-D-glucosamine oligomers and chitinous structures, whether from monocotyledonous or dicotyledonous plants (*Datura, Cucuma*, etc) or even from potato tubers, are located externally to seed structures. As chitin is a common constituent of cell walls in fungi and algae, such an external location of lectins on sensitive plant structures at a vulnerable stage of development is wholly compatible with the proposed defensive role. It is also very suggestive that a novel, lectin-like hydroxyproline-containing glycoprotein, with specificity towards *N*-acetyl-D-glucosamine oligomers, is synthesized in suspension cultures of *Phaseolus vulgaris* in response to fungal elicitors. This lectin is quite different from the usual seed lectin, PHA. It is synthesized initially on endomembranous structures and is then rapidly transported to cell walls, where it becomes progressively less extractable. Therefore, this lectin resembles those found in Solanaceae and is coinduced with enzymes of phytoalexin synthesis. On the basis of these findings, it has been proposed that the lectin is involved in the natural disease resistance of the bean plant (Bolwell, 1987).

Vegetative parts

Lectins are also found in the vegetative parts of plants. Although their concentration in these parts is usually, but not always, less than in seeds, these lectins are not necessarily less important than seed lectins. Indeed, more recent developments in unravelling the structure, location and functions of vegetative lectins have brought us closer to understanding the overall role(s) of lectins in the life-cycle of plants.

A number of lectins found in vegetative parts of plants are identical with those found in seeds. However, several lectins have been isolated from roots, leaves, barks, flowers, bulbs, rhizomes or other vegetative parts which differ from the seed lectins.

In one of the most comprehensive series of studies, Etzler and her colleagues have initially described the presence of at least three different lectins in various parts of the *Dolichos biflorus* plant. These studies have established that the seeds of this plant contain a well-characterized, *N*-

acetyl-D-galactosamine-specific and blood-group A-specific lectin (Etzler & Kabat, 1970; Hammarström *et al.*, 1977). In leaves and stems there is a somewhat different, but immunochemically cross-reactive (CRM) lectin, with sugar-specificity somewhat broader but similar to that of the seed lectin (Talbot & Etzler, 1978; Etzler & Borrebaeck, 1980). Roots, on the other hand, contained an entirely different lectin in which the amino acid composition, molecular weight and isoelectric point differed from the seed lectin, whilst the carbohydrate specificity of the root lectin remained the same as that of the seed lectin (Quinn & Etzler, 1987). Moreover, the root lectin docs not cross-react with antibodies raised against the native seed lectin, although it gives a weak reaction with antiserum obtained against denatured seed lectin preparations (Quinn & Etzler, 1987). It appears that lectins found in the seeds or roots of *Dolichos biflorus* are products of different genes, but the seed lectin and the CRM obtained from leaves and stems may be coded by a common gene and differences between them may arise from different post-translational modifications.

In more recent work from Etzler's laboratory, the construction, isolation and sequence of cDNAs for both the seed lectin and the DB58, leaf and stem lectin (previously called CRM), have been reported. The cDNAs for both the seed lectin and DB58 lectin code for polypeptide chains of 275 amino acids including 22 amino acids constituting the amino terminal signal sequences (Schnell *et al.*, 1987; Schnell & Etzler, 1987; 1988). In primary peptide sequence, the DB58 lectin shows 84% homology with that of the seed lectin and at the nucleotide level there is 92% homology. The seed lectin contains one glycosylation sequence, Asn–Asn–Ser, between residues 114–116, while there are two such sites in DB58, one Asn–Ser–Ser at residues 12–14 and one Asn–Lys–Ser at positions 79–81. Although glycosylation does not appear to be essential for the biological activity of lectins, such differences in glycosylation may conceivably affect the binding properties of these two lectins. However, it has also become clear that there are two separate mRNAs for the two *Dolichos biflorus* lectins, confirming that DB58 and the seed lectin are encoded by two separate and differentially expressed genes. Each gene encodes only one single polypeptide precursor. In the seed lectin, the two different types of subunits arise by differential proteolytic processing of the precursor and these then form a tetramer in the native lectin (Schnell *et al.*, 1987; Schnell & Etzler, 1987). In DB58, the cDNA represents a mRNA encoding the larger polypeptide of $M_r = 29,545$, which contains a 22 amino acid signal sequence. In the DB58 heterodimer the two different subunits are again the products of differential proteolytic processing of the single precursor polypeptide (Schnell & Etzler, 1988).

In addition to the close homology between the lectins of the seed and DB58 from *Dolichos biflorus*, a comparison of their sequences with other legume seed lectins indicates a high degree of structural conservation during evolution.

A similar situation has been found to exist in the perennial pasture legume, *Lotononis bainesii*. The roots of this plant contain a tetrameric glycoprotein lectin of $M_r = 118,000$ whose haemagglutination activity can be best inhibited by a number of disaccharides containing terminal non-reducing galactose residues. However, the lectin has no strict anomeric specificity and cannot distinguish between α- or β-linked disaccharides, nor is the identity of the terminal reducing sugar decisive. The root lectin has been purified to homogeneity by affinity chromatography but the corresponding seed lectin has only been partially purified because it is not absorbed by the usual galactose-specific lectin affinity reagents, although it is inhibited by galactose or galactono-1,4-lactone. The two lectins are immunochemically not related and neither the seed, nor the root lectins contain CRMs to either of them (Law & Strijdom, 1984*a*). The root lectin has been found to be localized in the tips of developing root hairs, which, as previously suggested for other lectins, may be favourable for host-plant recognition of infective strains of rhizobial bacteria (Law & Strijdom, 1984*b*).

In the leguminous tree, *Sophora japonica*, the seed lectin is a galactose-specific, tetrameric glycoprotein agglutinin of about 120,000 molecular weight. Its primary amino acid sequence is homologous to other legume lectins (Poretz *et al.*, 1974; Strosberg *et al.*, 1983). The leaves contain two closely related glycoprotein lectins which are not immunochemically identical to that from the seeds. These lectins differ in molecular weight and carbohydrate content but their carbohydrate specificities are virtually indistinguishable (Hankins, Kindinger & Shannon, 1987). A similar lectin has also been isolated by affinity chromatography on lactose–Sepharose-6B from the leaves of *Sophora japonica* (Ito, 1986). Its subunit weight of 40,000 is larger than either of the two leaf lectins isolated by Shannon and his colleagues, but its carbohydrate-specificity is the same. Another related lectin has been isolated from the flowers of the tree which contains two dissimilar subunits of $M_r = 40,000$ and 34,000 (Ito, 1986), yet displays the same sugar specificities as all the other lectins purified from different parts of the tree.

In more recent studies, the isolation of five similar *N*-acetyl-D-galactosamine-specific lectins from the bark of the *Sophora japonica* tree has been reported (Hankins, Kindinger & Shannon, 1988). They differ from each other or the seed and leaf lectins. Although their carbohydrate-specificities and haemagglutinating activities are indistinguishable at, or above, pH 8.5, they are appreciably different below this pH value. The bark lectins are tetrameric glycoproteins with different amino terminal amino acid sequences. They are made up of subunits of M_r = 30,000, 30,100 or 33,000 in different combinations which, somewhat unusually, are cross-linked by disulphide bonds, similar to that found with the lectin from the seeds of *Phaseolus lunatus*. It has been suggested that the subunit of M_r = 30,100 is a minor one and that the five isolectins are the result of a random combination of the 30,000 and 33,000 molecular weight subunits. Preliminary results indicate that most, if not all, *Sophora japonica* lectins are the products of related, but distinct genes (Hankins *et al.*, 1988). By specific immunogold labelling both leaf and bark lectins have been found to be exclusively sequestered in protein-filled storage vacuoles, which closely resemble the protein bodies found in the cotyledonary cells of seeds. Thus, despite their multiplicity and tissue specificity, all lectins in *Sophora japonica* are located in storage organelles (Herman, Hankins & Shannon, 1988).

Further interesting relationships between seed and root lectins have been revealed by studies with soyabean (*Glycine max*) roots. In this plant, all the lectins are closely related (Gade *et al.*, 1981). When the roots are stripped of all extractable lectin activity and then incubated for 15 h, lectin activity can again be demonstrated although the reappearance of the activity is not the result of *de novo* synthesis. As the properties of the reemerging lectin are similar to those of the seed lectin (Gade, Schmidt & Wold, 1983), its appearance suggests that it may have originated from a stored pool of previously synthesized lectin in axial tissues, such as exists in *Phaseolus vulgaris* (Manen & Pusztai, 1982).

In several plants, such as peas or beans, the lectins from the seeds and roots are closely similar, if not identical. For example, the presence of a lectin in pea roots similar to that in the seeds has been demonstrated by indirect immunofluorescence with antibodies raised to the seed lectin (Kato, Maruyama & Nakamura, 1981). Moreover, the lectin isolated from the roots was similar in most of its properties, including its specificity, to the seed lectin (Hosselet *et al.*, 1983). In contrast, the pea root lectin isolated by Gatehouse and Boulter (1980) was apparently somewhat different from the seed lectin. The presence of a lectin on the surface of root cells of *Pisum sativum* has been confirmed by a sensitive, enzyme-linked immunoassay (Diaz *et al.*, 1984) and the temporal changes in lectin concentration at different locations in the roots have been correlated in this way with the extent of *Rhizobia* infection (Diaz *et al.*, 1986). The final proof of the close structural relationship between the seed and root lectins in peas has been obtained by using the copy DNA of the seed lectin as a probe. In root tissue at 21 days, a mRNA of the same size as that found in the seeds was shown to be present although its concentration was about 4000 times less than that in the seeds. The lectin mRNA appears in the roots as early as the fourth day after sowing whilst the highest levels are obtained after about 10 days and just before nodulation (Buffard, Kaminski & Strosberg, 1988).

All vegetative tissues of kidney bean (*Phaseolus vulgaris*) contain lectins with similar properties and immunochemical reactivity (Borrebaeck, 1984; Castresana *et al.*, 1987). However, their concentration changes during the growth of the plant. It decreases sharply in the first weeks, reaches a minimum at the trifoliate leaf-stage and then increases steadily in the following phases of plant development (Castresana *et al.*, 1987). Definite proof of the presence of a lectin in root tissues of kidney bean, with similar properties to the seed lectin, PHA, has also been demonstrated by isolating small quantities of such a lectin from nodulated root tissues of a bean variety, Pinto III, hitherto regarded as lectin-free (Pusztai *et al.*, 1981*a*, 1982*a*). Although it is now generally accepted that Pinto III seeds contain very small amounts of a type-3, dimer lectin, different from PHA (Pusztai *et al.*, 1981*a*), the lectin synthesized by the roots of the Pinto III plant is much more like the generally occurring type-1 or type-2 seed PHA lectins (Pusztai *et al.*, 1981*a*, 1982*a*, 1983*a*). Moreover, the amounts of lectins in the roots increased appreciably on infection with *Rhizobium phaseolarum*. In accordance with this, the lectin content of bean roots was found to be low in the absence of the bacteria. This was in contrast to that found with pea roots, where the lectin mRNA concentration was not influenced by *Rhizobia* infection (Buffard *et al.*, 1988).

In the last few years it has become clear that lectins are more generally distributed in plant tissues than had been envisaged. In families other than Leguminoseae, it is not always the seeds which contain the greatest concentration of lectins. For example, in tulips (*Tulipa gesneriana*), a typical representative of the monocotyledonous plant family, Liliaceae, relatively large amounts of lectins are found in the bulbs. From these a tetramer lectin of $M_r = 67,000$ has been isolated by affinity chromatography on mannan–Sepharose-4B (Oda & Minami, 1986). This agglutinates yeast (*Saccharomyces cerevisiae*) cells but is inactive with other bacteria or animal erythrocytes. The specific inhibition of this lectin requires the presence of a hydroxyl group at C-2 of D-mannose. The lectin also recognizes (1–6)-linked manno-oligosaccharide units larger than mannobiose.

Another lectin, different from the D-mannose-specific yeast agglutinin, has also been purified by affinity chromatography on thyroglobulin–Sepharose-4B from tulip bulbs. This lectin has a molecular weight of 40,000 and contains two different subunits of $M_r = 26,000$ and 14,000; it specifically agglutinates rat and mouse erythrocytes, but has no effect on red cells from other animal species or on yeast cells (Oda *et al.*, 1987). The haemagglutination activity of the lectin cannot be inhibited by simple sugars or by a number of glycopeptides (Oda *et al.*, 1987).

Still another lectin, apparently different from the previous two bulb lectins isolated by the Japanese workers, has been purified by affinity chromatography on fetuin–Sepharose (Cammue, Peeters & Peumans, 1986). This is a much larger molecule, with $M_r = 120,000$, than the other two bulb lectins. Rather interestingly, although it exhibits complex sugar specificity when tested with rabbit erythrocytes, the agglutination of human cells by this high molecular weight bulb lectin is inhibited by *N*-acetyl-D-galactosamine, lactose, fucose and galactose (Cammue *et al.*, 1986).

The bulbs of snowdrops (*Galanthus nivalis*), a member of the monocotyledonous family of Amaryllidaceae, have also been shown to contain appreciable amounts of an interesting lectin, which recognizes D-mannose exclusively (van Damme, Allen & Peumans, 1987*a*). This is a tetramer lectin, composed of four identical subunits of $M_r = 13,000$. It readily agglutinates rabbit erythrocytes but it is completely inactive with human red cells (van Damme *et al.*, 1987*a*). Moreover, bulbs from three species of the same family of Amaryllidaceae, *Narcissus pseudonarcissus*, *Leucojum aestivum* and *Leucojum vernum*, have also been shown to possess D-mannose-specific lectins. These are serologically identical with that from snowdrops, and in strong contrast to other D-mannose-specific lectins, such as those from Vicieae, cannot be inhibited by D-glucose. Although their molecular structures are different, all lectins isolated so far contain the same subunit of $M_r = 13,000$. However, with the exception of the tetramer snowdrop lectin, the other lectins are dimers (van Damme, Allen & Peumans, 1988*a*). Rather interestingly, the leaves of the orchid tway blade (*Listera ovata*) also contain a similar, mannose-specific lectin (van Damme *et al.*, 1987*b*).

An unusual lectin, with specificity for *N*-acetyl-D-glucosamine oligomers, has been isolated from rhizomes of the stinging nettle (*Urtica dioica*). It is a relatively small lectin of $M_r = 8500$. In fact, it has been claimed to be the smallest phytohaemagglutinin found in plants so far (Peumans, De Ley & Broekert, 1984). The single chain polypeptide lectin is composed of 77 amino acid residues and contains no covalently bound carbohydrate. Although its haemagglutinating activity is low, the lectin interacts with a number of cells, including lymphocytes in which it induces the production of interferon (Peumans *et al.*, 1984). More recent work has revealed that the rhizomes contain a complex mixture of isolectins with very similar properties and specificities (van Damme & Peumans, 1987). However, the isolectins differ in amino acid composition. It has been suggested that the peptide sequences of the isolectins are encoded for by a family of closely related lectin genes (van Damme *et al.*, 1988*b*).

The rhizomes of ground elder (*Aegopodium podagraria*) also contain a lectin (Peumans *et al.*, 1985) which is very different from the nettle lectin. The elder rhizome lectin is specific for *N*-acetyl-D-galactosamine and has an unusually high molecular weight of about 480,000. It is probably an octamer comprising two different subunits. It is a typical non-seed lectin occurring exclusively in the underground rhizome, where it amounts to about 5% of the total protein. This concentration is different at various locations in the rhizomes and changes dramatically according to seasons; it is highest in winter months and lowest in the spring (Peumans *et al.*, 1985).

Several genera of Euphorbiaceae are known to be good sources for lectins. The best known and most studied lectin in plants of this genus occurs in the *Ricinus communis* seed. However, lectins in Euphorbiaceae are not confined to seeds only. For example, both the seeds (McPherson & Hoover, 1979; Falasca *et al.*, 1980) and the latex (Barbieri *et al.*, 1983) in *Hura crepitans* contain lectins. In a more comprehensive study, a wide range of agglutinating activities have been found in crude latex sera from 17 members of the genus *Euphorbia* (Lynn & Clevette-Radford, 1986). All these lectins have similar carbohydrate specificities, with lactose as the best inhibitor of their haemagglutination activity. Homogeneous lectin preparations have been isolated by affinity chromatography on lactose–agarose gels from the latex of seven species of the genus. Molecular weight values of the lectins vary from 60,000 to about 67,000, with subunit molecular weights between 27,000 and 38,000. All seven lectins contain covalently bound carbohydrates. It is, however, not clear what relationship, if any, these seven have with the much larger molecular weight (M_r = 140,000) glycoprotein lectins isolated by affinity chromatography on fetuin-agarose from the latex of three different species of *Euphorbia* (Nsimba-Lubaki, Allen & Peumans, 1986*a*). On the other hand, it is quite clear that the immunochemically related latex lectins, with specificity for D-galactose/*N*-acetyl-D-galactosamine, have very little in common with those isolated from Euphorbiaceae seeds and located in the endosperm (*Ricinus communis*) or in the embryo (*Hura* *crepitans*). The amounts of the lectins found in the latex amount to as much as 5% of the total protein. Obviously, this indicates that they may have important biological function(s) in the plant. Unfortunately their role is unknown so far.

Lectins are also found in the bark tissues of several tree species and some of these have been isolated and purified. For example, relatively large amounts of lectins have been obtained from the barks of elder (*Sambucus nigra*; Broekaert *et al.*, 1984), black locust (*Robinia pseudoacacia*; Horejsi, Haskovec & Kocourek, 1978) and golden chain (*Laburnum anagyroides*; Lutsik & Antonyuk, 1982). More recently (Nsimba-Lubaki *et al.*, 1986*b*) lectins have been purified by affinity chromatography on fetuin–Sepharose-4B from three different members of the genus *Sambucus*. All three are glycoproteins of similar amino acid composition, with molecular weight values of about 140,000 and best inhibited by *N*-acetyl-D-galactosamine, galactose and galactose-containing oligosaccharides (Nsimba-Lubaki *et al.*, 1986*b*). The amounts of both *Sambucus nigra* and *Robinia pseudoacacia* lectins are subject to considerable seasonal fluctuations. Large amounts accumulate in the protein bodies of the phloem parenchyma of bark tissues in these trees in the autumn, which are then depleted during new growth in springtime (Greenwood *et al.*, 1986). Thus, as barks serve as a form of vegetative storage tissue, the results suggest that bark lectins may be typical storage proteins (Nsimba-Lubaki & Peumans, 1986). As mentioned previously, barks of the tree, *Sophora japonica*, also contain lectins (Hankins *et al.*, 1988) and these have been clearly shown to be located in storage vacuoles of bark tissues (Herman *et al.*, 1988).

Lectins are also found in phloem exudates, such as from cucumber (Allen, 1979) and vegetable marrow (Sabnis & Hart, 1978). In more recent studies (Smith, Sabnis & Johnson, 1987) by using peroxidase-labelled antibodies and protein A-colloidal gold, the lectins have been localized in P-protein aggregates of sieve elements and companion cells in the tissues of *Cucurbita maxima* and cucurbit species. The lectin is, in fact, a component of the P-protein filaments and is associated with the sieve-tube reticulum lining the plasmalemma and it appears that it may serve as an anchorage point for the P-protein filaments in the parietal layer of sieve elements (Smith *et al.*, 1987).

Biosynthesis of plant lectins

A great deal of compelling evidence has accumulated from careful studies carried out in the last 20 years that, at least from the point of view of their biosynthesis, seed lectins resemble and behave like storage proteins.

Indeed, well-recognized storage proteins and lectins appear at about the same time during development in most legume seeds (Pusztai *et al.*, 1983*a*). Extensive studies have detailed the actual mechanism and location of the translation process in lectin biosynthesis. The route of subsequent transport of the primary product of translation, the occurrence of potential proteolytic and other processing steps, including reactions of glycosylation and de-glycosylation which may take place in the various intracellular organs *en route* and the final targeting and transport to the protein bodies has been established in detail from extensive studies. For example, the major steps in lectin biosynthesis in the seed cotyledons of kidney bean (*Phaseolus vulgaris*) have been unravelled by Chrispeels and his colleagues (Chrispeels, 1984). Accordingly, PHA is synthesized on the rough endoplasmic reticulum. The primary product is first transiently associated with membranes of endoplasmic reticulum. However, PHA extracted with detergents from these membranes has the same sedimentation constant and is as active a haemagglutinin as the seed lectin (Chrispeels & Bollini, 1982). Biosynthesis of PHA on polysomes attached to the endoplasmic reticulum includes the co-translational attachment of a high-mannose oligosaccharide chain via a dolicholphosphate intermediate (Bollini, Vitale & Chrispeels, 1983). The next step in the biosynthetic process is probably similar to that found with the biosynthesis of pea lectins. Accordingly, mRNAs for the PHA lectin may also include coding sequences for a hydrophobic amino terminal signal peptide similar to those found with the pea lectin (Higgins *et al.*, 1983*a*, *b*; Chrispeels, 1984). This sequence is thought to be responsible for the cotranslational transfer of the polypeptide chain from the membranes into the lumen of the endoplasmic reticulum by a mechanism similar to that which occurs with secretory proteins in mammalian systems (Walter, Gilmore & Blobel, 1984). During the transport of the newly synthesized lectin, a membrane-bound specific peptidase removes the signal peptide from the amino terminal, revealing the proper amino terminal amino acid of the native lectin (Higgins *et al.*, 1983*a*,*b*).

In the next step, *en route* to the protein bodies, the lectin is first processed through the Golgi membranes. Thus, on incubation with radioactively labelled fucose, the radioactive tracer-containing PHA (Chrispeels, 1983*a*) is found to be attached to organelles with density values of 1.13 and 1.22 g cm^{-3} in sucrose-density centrifugation experiments. Moreover, the results of pulse-chase experiments suggest that the fucose-labelled PHA is first associated with the Golgi membranes of average density of 1.13. From these it passes on to small electron-dense organelles, which have an average density of 1.22 g cm^{-3}. On the basis of protein composition, the electron-dense subcellular organelles are probably identical with initial forms of small protein bodies. The process of transport out of the Golgi apparatus can be effectively blocked by monensin. In the presence of this ionophore, the transfer of PHA to the protein bodies does not take place (Chrispeels, 1983*b*; Stinissen, Peumans & Chrispeels, 1985). During transport through the Golgi membranes, other changes in the oligosaccharide side-chains of the lectin occur. For example, a residue of *N*-acetyl-D-glucosamine becomes transiently attached; this is finally removed by a protein-body associated *β-N*-acetyl-D-glucosaminidase after the PHA reaches the protein bodies (Vitale & Chrispeels, 1984). As apparently glycosylation is not an obligatory step in the intracellular transport of PHA (Bollini *et al.*, 1985), it is unclear why these complex glycosylation and de-glycosylation reactions occur. For example, glycosylation of PHA in excised developing cotyledons can be prevented in the presence of tunicamycin. Despite this, the non-glycosylated lectin still reaches the final site of its normal deposition, i.e. the protein bodies, where it is fully active both as mitogen and haemagglutinin (Bollini *et al.*, 1985). Thus, the carbohydrate side-chain in PHA is non-essential for full biological activity, in full agreement with the more general observation that the oligosaccharide side-chains of most viral and animal glycoproteins are not required for biological activity (Olden, Parent & White, 1982).

The rather intriguing finding that residue(s) of *N*-acetyl-D-glucosamine is attached to PHA in the Golgi apparatus *en route* to the protein bodies, has been explored in more detail recently. It is known that, after deposition in the protein bodies, these terminal *N*-acetyl-D-glucosamine residues are removed (Vitale & Chrispeels, 1984) and it has been suggested that the temporary attachment of such terminal sugar residues to PHA may serve as a targeting aid to guide the lectin to its final destination. Indeed, after [^{14}C]-glucosamine labelling of developing pea seed cotyledons, the subcellular structures of the endoplasmic reticulum and the Golgi apparatus and an organelle with a density of about 1.22 g cm^{-3}, probably a transit vesicle, have all incorporated the radioactive label. Moreover, on lysis, a number of proteins, with terminal *N*-acetyl-D-glucosamine residues, is released from the matrices of these organelles. Membrane proteins which also contain terminal *N*-acetyl-D-glucosamine residues, can be specifically released from the membranes after the removal of the labelled matrix proteins by washing with solutions of *N*-acetyl-D-glucosamine. Of these, one such glycoprotein containing terminal *N*-acetyl-D-glucosamine of $M_r = 67,000$ has been partially purified and shown to be present in all three organelles and also in protein body membranes. This membrane glycoprotein, with transiently attached *N*-acetyl-D-glucosamine residues, is thought to be a receptor protein component (lectin) of the membranes of the developing pea cotyledons and may be involved in directing protein trafficking from one compartment to another by using the terminal aminosugar residues as targeting ligands (Harley & Beevers, 1988).

The existence of a *Phaseolus vulgaris* cultivar, Pinto III, in which the small amount of constituent type-3 lectin is different from the usual kidney bean seed PHA isolectins (Pusztai *et al.*, 1981*a*, 1982*a*), has facilitated studies of the transcriptional and post-transcriptional controls of lectin-gene expression in developing cotyledons. Although this cultivar contains mRNA specific for the usual PHA isolectins, its amount and the corresponding translational activity are only about 1% of those found in the more generally occurring *Phaseolus vulgaris* varieties. This severe reduction in the amounts of mRNA specific for type-1 and type-2 PHA lectins occurs despite that all cultivars, including Pinto III, contain the same or similar PHA isolectin genes without any major structural rearrangements. Accordingly, in Pinto III seeds, the transcription rate of the DNA must be deficient in agreement with experimental findings which show that in Pinto III the transcription rate of the PHA gene sequences is only about 20% of that found in normal, high-lectin kidney bean cultivars. In addition, the accumulation of only small amounts of PHA mRNA may indicate that this mRNA is inherently unstable in the lectin-deficient Pinto III seeds (Chappel & Chrispeels, 1986).

In general outlines, the biosynthesis of other lectins may be very similar to what has been found for PHA isolectins. If there are any differences, they are likely to occur in the various post-translational modification steps. For example, heterodimer legume lectins are usually synthesized first as a single pre-prolectin polypeptide. The additional proteolytic processing steps required to produce the α- and β-subunits of two-chain lectins are known to occur after the signal sequence is removed and the polypeptide is transferred into the lumen of the endoplasmic reticulum *en route* to the protein bodies (Higgins *et al.*, 1983*b*). With pea lectin, which is transported as a prolectin, the splitting of the single chain to produce the two-chain mature pea lectin takes place in the protein bodies (Higgins *et al.*, 1983*a*). Interestingly, the scission of the single polypeptide chain in both pea and *Vicia faba* prolectins occurs after an asparagine residue in the peptide sequence (Hemperley & Cunningham, 1983).

An even more striking and different post-translational modification is known to occur during the biosynthesis of concanavalin A (Fig. 1.1). As has been mentioned already, the midchain splitting of the prolectin is followed by a transposition and ligation step (Carrington *et al.*, 1985; Bowles & Pappin, 1988). Incomplete ligation of the broken fragments may also explain the existence of appreciable amounts of two fragments which contain residues of 1 to 118 and 119–237. These are found in all concanavalin A preparations isolated from the seeds of jack bean (*Canavalia ensiformis*) even when the extent of proteolytic breakdown during purification is minimized.

With recent developments it is becoming increasingly clear that seed lectin gene expression is developmentally regulated. Thus, transcripts of seed lectin genes may be almost entirely absent, or present in small amounts only, in vegetative tissues. In contrast, in the developing seed the transcription rate increases rapidly up to about midway through development, after which it declines with equal rapidity. Lectins appear in the developing seed at a precisely regulated time interval after anthesis.

Lectin biosynthesis and its genetic control are topics of great current interest for geneticists. With the aid of recombinant DNA technology, lectin genes and their transcription have been studied in many plants, including *Phaseolus vulgaris* (Hoffman *et al.*, 1982), soyabean (Goldberg, Hoschek & Vodkin, 1983) and peas (Higgins *et al.*, 1983*b*). These studies have confirmed the expected patterns in the increase of mRNA concentrations with seed development and sequence analysis of cloned lectin cDNAs has supported previous findings that all lectins are synthesized as signal sequence-containing prolectins, ready for transfer into the lumen of the endoplasmic reticulum.

The low level of seed lectin biosynthesis in the lectin-defective, Pinto III, kidney bean cultivar is the result of defective transcription rates (Chappel & Chrispeels, 1986). Similarly, lectin gene transcription is at an appreciably reduced level in the soyabean mutant line which produces no seed lectin (L⁻). Both L^+ and L^- soyabean lines contain two lectin genes, but, in the lectin-defective mutant, gene expression is blocked by a large, 3.4 kb DNA sequence insertion (Vodkin *et al.*, 1983). The reduced transcription rate in the mutant gene is due to the presence of this transposable element in the DNA. In contrast, the reduced rate of transcription in Pinto III is apparently not due to the presence of such a transposon.

Biosynthesis of the lectin in castor bean (*Ricinus communis*), the most thoroughly studied plant in Euphorbiaceae, occurs by a mechanism similar to that described for the synthesis of two-chain legume lectins. The nucleotide sequence of the lectin cDNA indicates that the amino acid sequence of the initial product of biosynthesis starts with a signal peptide containing 24 amino acid residues joined to the amino terminal of the A-chain, which is linked to the B-chain by a peptide sequence containing 12 amino acids (Lamb *et al.*, 1985). As has been found with other lectins, the pre-prolectin is synthesized on polysomes attached to membranes of the endoplasmic reticulum. Next, during the transference of the prolectin into the lumen of the endoplasmic reticulum, the signal peptide is removed and the polypeptide chain is simultaneously polymannosylated (Roberts & Lord, 1981). The glycoprotein prolectin is then transported to the Golgi apparatus, where the oligosaccharide chains are subjected to a number of modification reactions in the membranes and vesicles of the organ (Lord, 1985*a,b*). Finally, the prolectin is deposited in protein bodies and, by proteolytically removing the linking peptide sequence (Harley & Lord, 1985), it is converted into the final product of native lectin, RCA_{II}, containing separate A and B polypeptide chains. It has also been established that the respective subunits of RCA_I and RCA_{II} are products of closely related, but distinct genes of a multigene family.

The biosynthesis and processing of rice and cereal lectins (Stinissen, Peumans & Carlier, 1982, 1983) is also known to be a two-step process. These lectins are synthesized as high molecular weight (23,000) pro-lectins initially and then post-translationally transformed by proteolysis into subunits of $M_r = 18,000$. Finally, the rice lectin is further cleaved producing two smaller subunits, with M_r values of about 10,000 and 8000.

The *in vitro* translation product of polyadenylated RNA isolated from developing wheat embryos is a lectin polypeptide of $M_r = 21,000$, but the prolectin synthesized *in vivo* is considerably larger, $M_r = 23,000$. If the developing embryos are exposed to $[^3H]$ mannose, this *in vivo* prolectin is radioactively labelled. In contrast, the *in vivo* product of biosynthesis in the presence of tunicamycin is a prolectin of $M_r = 20,000$ and not 23,000. These results have been interpreted to mean that the pre-prolectin, which is synthesized *in vivo*, is co-translationally processed by the removal of the signal peptide sequence and by the addition of a glycan moiety (Mansfield *et al.*, 1988). The amino acid sequence deduced from the nucleotide sequence of a *c*DNA clone has indicated that, rather unusually, this prolectin contains a 15 amino acid-containing extension attached to the known carboxyl terminal amino acid of wheatgerm agglutinin. This extension sequence also contains a site for *N*-linked glycosylation of the prolectin (Raikhel & Wilkins, 1987). The glycosyl-ated precursor is finally post-translationally processed to the mature form of wheatgerm agglutinin by proteolytic cleavage and subsequent removal of the glycosylated carboxyl terminal extension peptide (Mansfield *et al.*,

1988). The use of the cDNA clone as a hybridization probe to encode the wheatgerm agglutinin precursor has shown that the level of wheatgerm agglutinin-specific mRNA can be modulated by treatment of excised wheat embryos with abscisic acid (Raikhel & Wilkins, 1987). This finding is a rather elegant confirmation of a previous observation made with wheat plants that the concentration of wheatgerm agglutinin increases in the embryos in the presence of abscisic acid (Triplett & Quatrano, 1982).

Wheatgerm agglutinin and chitinase may both function as anti-microbial agents (Mirelman *et al.*, 1975). Indeed, the amino terminal amino acid sequence of chitinase (Broglie, Gaynor & Broglie, 1986) and that of the four regions of the wheatgerm agglutinin molecule show appreciable homology. Moreover, the amino acid residues implicated in the primary carbohydrate-binding site of the lectin (Wright, 1984) are all found in this region of homology (Raikhel & Wilkins, 1987). Thus, the homologous peptide sequence found in the two chitin-binding proteins of wheatgerms may have some functional significance in antimicrobial activity.

In conclusion, it is now generally accepted that lectins are present in all tissues of plants. It has become clear from differences in their amounts in various parts of plants and changes in their concentration, molecular properties and subcellular localization during the life-cycle of plants that lectins are probably multifunctional proteins. Both the high degree of structural homology and the close developmental control of gene expression argue strongly that lectins may have important, although so far unknown, function(s) for plants.

4

Biological functions in plants

Interactions with endogenous receptors

The existence of functional endogenous receptors for lectins in plants has often been suggested. Such receptors, by definition, are good inhibitors of lectins from the same plant and therefore their presence ought to be relatively easy to demonstrate. Indeed, the occurrence of endogenous inhibitors in the seeds of *Vicia cracca* had been suggested as early as 1960 (Renkonnen, 1960) and similar observations have been made for the seeds of *Dioclea sclerocarpa* (Pusztai & Moreira, unpublished observations).

Most legume seeds contain appreciable amounts of α-galactosidic oligosaccharides, such as raffinose, stachyose and verbascose, which are good inhibitors of galactose-specific lectins. In seeds such as soyabean, which contain both such lectins and also α-galactosides, the oligosaccharides may be regarded as natural lectin-binders. The lectin, which is a major example of the N-acetyl-D-galactosamine/galactose-specific class of lectins co-exists with raffinose-type oligosaccharides in the seeds. These naturally occurring sugars are about 14% as effective in inhibiting the haemagglutination activity of the lectin as N-acetyl-D-galactosamine (Pereira *et al.*, 1974). It is thus not surprising that less than 20% of the lectin can be absorbed on to guar gum-based affinity absorbents from non-dialysed extracts of soyabean. However, both by dialysis and/or precipitation of the proteins with ammonium sulphate followed by dialysis, most of the natural lectin-binders are removed from the soya-bean extracts. The lectin in the extract is then selectively and quantitat-ively absorbed by the affinity column. Similar results can be obtained if the seed meal is first washed by 60% aqueous ethanol to remove the bulk of low molecular weight glycosides before the extraction of the lectin (Pusztai, Stewart & Watt, in press).

Endogenous receptors for lectins (lectin-binders) of larger molecular size have also been isolated from seeds by affinity supports containing lectins from the same plant. These seeds included *Pisum sativum, Canavalia ensiformis, Vicia faba, Vicia sativa* and *Ricinus communis* (Gansera, Schurz & Rüdiger, 1979; Gebauer *et al.*, 1979). Most receptors are, apparently, simple glycoproteins and binding to these natural receptors is inhibited by sugars for which the lectins are specific. The lectin-binders have no haemagglutinating activity, although some of them may possess strong mitogenic activity for peripheral lymphocytes (Gebauer, Schimpl & Rüdiger, 1982).

The precise nature of the interaction between lectin-binders and lectins is not clear. The demonstration that some of the interactions can be abolished by low concentrations of sodium chloride and not just by haptenic carbohydrates, indicates a lack of specificity which places these reactions outside the strict definition of lectins (Einhoff & Rüdiger, 1986*a,b*). However, interactions between lectins and other components of the seeds may still have great importance for plants, even if these are not based on the biological activity of the lectins.

Bowles and Marcus (1981) have also isolated lectin-receptors from extracts of soyabean or jackbean seeds by affinity chromatography on the appropriate immobilized lectins. However, their amounts are very small in comparison with the usually large amounts of lectins in the same seeds. This questions the physiological significance of the lectin-receptors. However, the observation that the lectin-receptor in jackbean is identical with the heavy subunit of seed α-mannosidase, may have potentially far-reaching consequences. It also suggests that such receptors may be involved in enzyme sequestration within intracellular structures of seed cells (Bowles, Andralojc & Marcus, 1982).

By applying a radioaffinity assay for the binding of the soyabean lectin, a wide range of seed lectin-binding components has been found in soyabean seeds, including several glycoconjugates and membrane components of both low and high molecular weights (Bond, Chaplin & Bowles, 1985). In order to clarify the relationship between the various concanavalin A-binding components, seed polypeptides with affinity for the lectin were studied simultaneously by two different affinity methods. By using $[^{125}I]$-concanavalin A, the heavy subunit of α-mannosidase is recognized with the overlay technique. However, the native enzyme does not bind to concanavalin A-Sepharose and only the denatured lectin binds to α-mannosidase. Moreover, jackbean extracts contain a single minor native protein which binds concanavalin A; this is different from α-mannosidase (Marcus, Maycox & Bowles, 1989) and therefore it is unlikely that concanavalin A receptors play a major role in enzyme sequestration in developing jackbean seeds.

Lectins from other seeds, such as PHA from kidney bean, may interact with membrane components of protein bodies in cotyledonary cells (Pusztai *et al.*, 1979*a*). Similarly, wheatgerm agglutinin binds to protein bodies and nucellar epidermis (Baldo, Boniface & Simmonds, 1982*a,b*) and the plasma membranes of soyabean root protoplasts also bind the seed lectin (Metcalf *et al.*, 1983).

In conclusion, on the basis of the scanty and sometimes contradictory evidence available, it is difficult to visualize the physiological significance of endogenous receptors in plant metabolism. It is quite possible that when plant tissues are disrupted by homogenization or extraction, seed proteins or glycoproteins may artificially bind some of the abundant seed glycolipids, phenolic or other glycosides, polyphenols or other metabolites also released from seed cells. Such components or their binding products may then react with seed lectins either through specific carbohydrate binding or by other, non-specific forces. It may be of particular significance that lectin-binders are usually present in seeds in small amounts and that some of their interactions with seed lectins may be abolished by changes in salt concentration rather than by displacement with carbohydrate haptens of the appropriate specificity. Moreover, at least for the time being, no morphological or ultrastructural supporting evidence exists to demonstrate the presence of lectin-receptors *in situ* in seed cells and even less evidence exists for the proposed *in vivo* binding of lectins to endogenous receptors in seeds.

Interactions with exogenous receptors

The wide distribution of lectins in all tissues of plants is now generally accepted. Some of these lectins are known to be located on the external and exposed surfaces of plant tissues. It is this location, in combination with their ready ability to distinguish between different cells containing different membrane carbohydrates which argues strongly in favour of a potentially important role for lectins in combating the invasion of plants by pathogenic organisms. Indeed, as antibodies and lectins have similarities in their biological effects, it has been suggested that lectins function as plant antibodies. However, such similarities are superficial and the idea has been abandoned; for all their possible role in the defence of plants against predators, lectins are not antibodies. There is no evidence at all to suggest that plants possess an immune system, particularly when comparisons are made with the highly complex and adaptable animal immune system (Clarke & Knox, 1979).

Defence against microorganisms and potential predators
It has been suggested that some plant lectins are particularly adapted to
fighting off attacks by fungi in whose cell wall chitin is a major component
(Mirelman *et al.*, 1975). By facing towards the external environment, all
chitin-binding lectins from monocotyledonous, dicotyledonous plants,
tubers, etc, and particularly the agglutinins from wheatgerm or other
cereals, are strategically well placed for such a role. Thus, fluorescein-
labelled wheatgerm agglutinin is attached to the hyphal tips and septa of
Trichoderma viride and the binding, as expected from the sugar-
specificity of the lectin, can be inhibited by chitotriose. Moreover,
wheatgerm agglutinin does not bind to the mature hyphae in which chitin
is not accessible (Mirelman *et al.*, 1975). As the lectin from wheatgerm
interferes with the biosynthesis of chitin, it is not surprising that it inhibits
spore germination and the growth of the fungus. The hyphal tips, septa
and young spores of all chitin-containing fungi are similarly sensitive to
the effects of wheatgerm agglutinin (Barkai-Golan, Mirelman & Sharon,
1978; Galun *et al.*, 1976). A somewhat similar binding to young spores
and mature hyphae of *Penicillium* and *Aspergillus* species has been
demonstrated for lectins from soyabean and peanut. These lectins have
growth inhibitory activities for some fungi and interfere with germination
of their spores (Barkai-Golan *et al.*, 1978); soyabean lectin also affects
mycelial growth of *Phytophthora megasperma* (Gibson *et al.*, 1982).
Potato lectin, another chitin-binding lectin, has similar effects on spore
germination and growth of hyphae of *Botrytis cinerea* (Callow, 1977).

Although these observations appear to support the suggested protective function of some plant lectins in defending plants against invading
pathogenic fungi, most of the evidence is indirect and circumstantial.
Plants are most vulnerable to such attacks during imbibition, germination
and seedling stage, i.e. before the plant is firmly established. To be
effective in their defence role, lectins must be localized in the right places
in the plant and be available in sufficient concentrations. However, the
evidence for such a role is indirect even in the case of the extensively
studied wheatgerm agglutinin. Although this lectin is found at sites where
pathogens usually attack, i.e. the external surfaces of the embryo (Mishkind *et al.*, 1982) and, in addition, relatively large amounts of lectins are
released into the surrounding environment from the imbibed seeds
(Fountain *et al.*, 1977; Hwang, Yang & Foard, 1978), direct experimental
evidence for the defence role of the lectin is still not available. The
evidence for other plants is even more indirect. In some plants, such as
the soyabean, those seeds which show high resistance to attack by
pathogenic fungi not only have lectins in a higher concentration but they

also release them earlier than do corresponding susceptible cultivars (Gibson *et al.*, 1982). Similarly, in *Datura stramonium* seeds, the lectin and other carbohydrate-binding proteins are found almost exclusively in the outer seed tissues of seed coats and epidermis (Broekaert *et al.*, 1988). Imbibition of mature seeds results in a rapid and specific release of the seed lectin and other carbohydrate-binding proteins into the rhizosphere. It is envisaged that in one of the initial steps in the attack by pathogens, the microorganisms first become attached to the cell wall or plasma-lemma of cells in the outer tissues of plants. Lectins localized on external surfaces of such tissues or those released into the rhizosphere, by specifically recognizing and binding to the cells of the invading micro-organism, can interfere with this attachment. Thus, the presence of lectins may provide a protective shield around the germinating seed and facilitate its survival at the most vulnerable stage of plant autotrophy.

In addition to lectins, which are constitutive components of seed cells with potential antifungal activities, plants can also respond to fungal injection by synthesizing a number of other defensive agents. Phytoalex-ins are the best known of such agents. Plants also synthesize and accumulate large amounts of hydroxyproline-rich cell wall glycoproteins, called extensins (Esquerre-Tugaye & Maxau, 1974). Such cell wall structural glycoproteins in some plants may include lectins, similar to those found in Solanaceae. The function of these lectins as defensive agents has been well established in wounded and infected plants (Leach, Cantrell & Sequeira, 1982). Although such cell wall lectins are probably related to extensins and other wall arabinogalactan proteins and result from plant responses to fungal elicitors, the two types of cell wall glycoproteins are both structurally and functionally different. Moreover, in suspension cultures of kidney bean, the synthesis of the cell wall-related lectin, the *Datura*-like hydroxyproline-rich glycoprotein lectin, precedes the synthesis of cell wall extensin (Bolwell, 1987). The obser-vation that the biosynthesis of the lectin occurs concomitantly with the synthesis of enzymes involved in phytoalexin production suggests that the lectin may play a role in disease resistance (Bolwell, 1987).

Lectins may protect plants against invading, harmful bacteria (Sequeira, 1978, 1984). By recognizing and binding to the saprophytic but not to the pathogenic species of bacteria, strategically localized lectins may confer resistance to the plant. Such a reaction has been described for potato lectin which binds only to saprophytic *Pseudomonas solana-cearum*. The argument has also been turned round by suggesting that this is precisely the reason why one species of the same bacterial strain is virulent while the other is not (Sequeira, Gaard & De Zoeten, 1977;

Sequeira & Graham, 1977). Similar observations have been made with the cell walls of leaves from *Phaseolus vulgaris*, which apparently bind only the saprophytic species of *Pseudomonas*. Although the kidney bean seed lectin agglutinates the saprophytic but not the pathogenic strains of the bacteria (Sing & Schroth, 1977), the true significance and general applicability of such findings for any proposed *in vivo* antibacterial role for lectins needs further detailed studies. It appears that the problems encountered when attempts are made to corroborate the suggested role of lectins in the defence of plants against microbial attacks are very similar to those discussed in connection with the putative functions of endogenous lectin-binders in the previous section (see p. 59). There is a clear need for more detailed morphological evidence not only about the localization of lectins in sensitive tissues of plants, but also for more information about the biochemical mechanism of interactions between surface lectins and glycoconjugates of both plant and invading bacteria. As yet, no direct experimental evidence exists to support the proposed protective function of lectins.

Very similar considerations may apply to other suggested roles of lectins, some of which actually runs counter to the proposed protective function. There is some evidence to indicate that some lectins in plants may actually promote pathogenesis (Sequeira, 1978) and the oft-quoted example for this is the binding of the toxin from the fungus, *Helmintho-sporium sacchari*, by a lectin-like membrane protein from sugarcane (Strobel, 1974). This membrane protein behaves like a lectin, it is multivalent and can be inhibited by α-galactosides. However, there are several inconsistencies regarding the structure of the toxin and its carbohydrate composition. Thus, the mechanism of its binding to the lectin-like membrane protein is still controversial and needs further corroboration before it can be accepted as a prime example of the promotion of pathogenesis by lectins.

Finally, lectins may serve as defensive agents against attack by beetles, insects and other predators, including even mammalian species (Janzen, 1981). For example, the larvae of the bruchid beetle is killed by incorporating kidney bean lectins in their diets (Janzen, Juster & Liener, 1976). Similarly, the incorporation of a highly purified preparation of soyabean lectin, at 1% concentration in the diet of the larvae of the insect *Manduca sexta*, inhibits their growth (Shukle & Murdoch, 1983). Further advances in our understanding of the mechanism of growth inhibition of insect larvae by lectins have been obtained by immunofluorescence studies. Thus, it has been shown by the use of monospecific anti-lectin antibodies that in the gut of the larvae of *Callosobruchus maculatus*, raised on artificial seeds containing PHA isolectins, the dietary lectin binds to the surface of the gut. This binding probably interferes with food absorption and leads to poor growth of the larvae (Gatehouse *et al.*, 1984). On the other hand, as no binding of PHA to midgut epithelial cells occurs in the bean weevil (*Acanthoscelides obtectus*) it suggests a plausible explanation for the observed lack of toxicity of PHA-containing bean seeds for the larvae of this major bean seed storage pest. Furthermore, although the reorganization of gut epithelium on pupation allows the lectin to diffuse into adipose tissue and to bind to surface membranes of fat cells, the disruptive effects of PHA are manifest only in *Callosobruchus maculatus*, for which PHA is toxic. In contrast, PHA does not bind to the fat cells of *Acanthoscelides obtectus*, which may explain why PHA is not toxic for this insect (Gatehouse *et al.*, 1989).

Further indirect evidence for such a protective role has been provided by a survey carried out on mature seeds of 59 species of tropical Leguminoseae (Janzen *et al.*, 1986). The results suggest that the most likely cause for a seed being rejected as food by a potential predator may be connected with the presence of so-called defensive compounds in the seed. There are potentially three or four classes of such compounds and some seeds may contain individual examples of the various classes or even a combination of some or all of them. The seed lectins provide prime examples of these defensive compounds (Janzen *et al.*, 1986).

There is also some indirect evidence to suggest that the resistance of a number of *Phaseolus vulgaris* cultivars to bruchid beetles, bean weevils, etc, is due to the presence in their seeds of a lectin-related protein, called arcelin (Schoonhoven, Cardona & Valor, 1983; Osborn, Burrow & Bliss, 1988). It is, however, not yet clear how this protein is related to the more usual PHA isolectins in kidney bean seeds and by what mechanism it confers resistance to bean seeds against predatory insects.

Thus, it appears that there is long way to go before the general occurrence of a lectin-based defensive system in plants can gain universal acceptance. Direct and detailed experimentation, along the lines provided by the immunofluorescence studies of Gatehouse and her colleagues (Gatehouse *et al.*, 1984, 1989) is needed to demonstrate the direct involvement of seed lectins in interactions with potential predators. For understanding what renders a seed resistant to attack and how this is accomplished, studies of the likely mechanism(s) of the interference by lectins in one or more of the vital life processes or functions of predators are also needed. Hopefully, with such an understanding, resistant seed cultivars can be provided at will for all potential uses by genetic engineering or even simply by genetic selection.

Legume–Rhizobium *symbiosis*

One of the potentially most important interactions of plant lectins with exogenous cellular receptors is the legume root–*Rhizobium*, host–symbiont relationship. Indeed, lectins have long been suggested to be implicated in symbiotic reactions (Hamblin & Kent, 1973). Fixation of atmospheric nitrogen by plants is of great agricultural importance and has been exploited by man for centuries. However, despite great effort and extensive experimental work, the precise nature of this interaction between the roots of legume plants and symbiotic microorganisms and the underlying biochemical reaction mechanism and genetics of the symbiosis are controversial and still not properly established. The great and general interest in the importance of symbiosis is clearly shown by the ever-increasing number of experimental papers and reviews published in the last decade (Dazzo & Truchet, 1983; Pusztai *et al.*, 1983*a*; Etzler, 1986; Halverson & Stacey, 1986*a*; Rolfe *et al.*, 1986; Stacey *et al.*, 1986; Kijne *et al.*, 1986). It appears, from a scientific consensus, that the experimental evidence obtained for the involvement of plant lectins in specific recognition and subsequent binding reactions between nodulating bacteria and the roots of host-plants, is strongest for the symbiotic interaction between clover (*Medicago sativa*) roots and its specific nodulating bacteria, *Rhizobium trifolii*. However, there is also an increasing body of evidence to suggest that similar processes may occur between plants, such as soyabean and peas, and their specific rhizobial bacteria during symbiotic interactions (Stacey *et al.*, 1986; Kijne *et al.*, 1986).

The process of bacterial nodulation is complex and, in successful infections, a great number of interlinked events occur. However, from the point of view of this review, the main and obvious concern is whether plant lectins are the main effective agents of cellular recognition, and, if so, what is the biochemical mechanism of the symbiotic reaction.

It is common knowledge that a very large number of bacteria exists in the rhizosphere of plant roots and that a proportion of these becomes attached to root hairs loosely and in a non-specific manner. However, as determined by the host, a small proportion of the microorganisms present binds to the root hairs more firmly and specifically. This is the first and prerequisite step for the success of the infection. As it is now visualized (Rolfe *et al.*, 1986), in one of the earliest events of the infection, the *Rhizobium* and the *nod* D gene carried by the microorganism respond to a number of hydroxyflavones secreted by the plant and subsequently the bacteria express an unknown number of further bacterial nodulation genes. The expression of the *nod* genes is required for the success of the host-specific nodulation of clover roots. The second important event in the infection process is the production of receptors on the surface of the *Rhizobium trifolii* cells which are required for the binding of the clover lectin, trifoliin A. It is known that the *nod* genes are deeply involved in the synthesis of such lectin-receptors of the bacteria.

In the first phase of the attachment, called the docking stage, the bacteria produce a fibrillar capsule which makes physical contact with the electron-dense globular aggregates found on the outer surface of the cell walls of root hair cells. It has been demonstrated experimentally that both the epidermal cell walls of clover roots and the bacterial surface contain a number of immunochemically cross-reacting glycoconjugates which may also function as receptors for the lectin, trifoliin A (Dazzo, Yanke & Brill, 1978; Dazzo & Brill, 1979). It is envisaged that the lectin binds in a hapten-reversible manner to both the root hairs and the bacteria and forms a bridge between them. This step can be inhibited specifically by the haptenic sugar, 2-deoxy-D-glucose and the attachment may be totally abolished in the presence of this sugar. In contrast, the same sugar is totally ineffective in other plant root–*Rhizobium* symbiotic systems. Binding studies with fluorescein isothiocyanate-labelled *Rhizobium trifolii* capsular polysaccharides have shown the existence of lectin-receptor sites on the root hairs. Moreover, the number of receptors found on the differentiated tissues of epidermal root surface is the highest on root tips.

In the second phase of the bacterial infection, a bacterial fibrillar material with adhesive properties anchors the bacterial cell to the surface of the root hair. This step is probably also necessary for inducing root hair curling and the penetration of the bacteria through the plant cell wall into root cells. For successful infection, however, a well-regulated interplay between the host and several bacterial host range genes is required. Non-homologous bacteria also contain several *nod* genes of *Rhizobium trifolii*, as these are functionally conserved in different species of the bacteria. However, the non-homologous bacteria either do not induce hair-curling or, even if they do, fail to infect the plant. Therefore, in non-homologous *Rhizobia*, the presence of incorrect host range genes elicit a rejective response from the clover root cells. It is, thus, quite clear that although the attachment of bacteria to the roots is a prerequisite step of symbiosis, it is far from being the only requirement for successful infection.

There is an additional complication in the infection process, as root hairs are known to be only transiently susceptible to infection. The expression of the lectin-receptor polysaccharide on the surface of the bacteria is controlled by the root hair-attachment (*roa*) genes, which are carried on large transmissible plasmids. Although these genes are transferable to non-infective mutants and these bacteria may even be transformed into lectin-binding mutants, those which are originally noninfective remain ineffective and the mutants do not induce root hairattachment.

Further studies have revealed that the plant is able to resist the invading bacteria by a hypersensitive rejection response even after the bacteria have successfully degraded the cell wall of root hair cells, penetrated the cells and managed to synthesize an infection thread. The existence of such positive or negative systemic plant responses to infection has recently come to light by split-root assays. When two roots on the same plant are infected with the same strain of bacteria at the same time, both roots form a roughly equal number of nodules. However, when the inoculation of the second root is delayed for over 24 h, the first root appears to inhibit the nodulation of the second root.

Studies described above have shown the great importance of postattachment events in successful symbiotic reactions. The initial lectinmediated binding of bacteria to root hair cells is now known to be a signal which induces the activation of a number of bacterial genes. As a result, the bacteria synthesize a number of specific products, which also include more lectin-binding cell wall polysaccharide antigens. In turn, these bacterial products act as signals and switch on plant-host cell genes to enable the plant cells to synthesize various enzymes and other factors which then react with bacterial surfaces. All these processes are directed for facilitating, in a positive and co-operative way, the enhancement of bacterial penetration, formation of infection thread and the success of nodulation. Thus, the lectin-mediated recognition and binding is but a first step in a complex exchange of signals between host and bacteria. Although, in their origin, all these events can be traced back to the initial lectin-mediated recognition step, signal-exchange is an ongoing process during the entire symbiotic attachment and successful infection and nodulation require conditions which favour a great number of interrelated but discrete steps (Halverson & Stacey, 1986a).

Despite the great advances in our understanding of the mechanism of *Rhizobium*–legume symbiosis, mainly due to the concentrated effort of Dazzo and his colleagues working with *Rhizobium trifolii*–clover symbiosis, the general applicability of this particular interaction to all symbiotic reactions has not yet been clearly established. As other symbiotic reactions have not been studied with the same intensity of effort, we know much less about them. Although recent studies with the soyabean root–*Bradyrhizobium japonicum* symbiotic system have given further support to the lectin-mediated bacterial recognition hypothesis (Stacey *et al.*, 1986), to gain further insights into the interaction mechanism, exploration of other experimental avenues is clearly needed.

One of the main complications and contradictions in the root–bacteria binding process, which has to be satisfactorily resolved, is that rhizobial bacteria which do not infect the roots of some legume species may still bind the lectin from the same legume seeds (Chen & Phillips, 1976). An explanation is also required for the converse observation that the lectin isolated from some legume seeds may not bind to all species of *Rhizobium* capable of infecting the same legume plant successfully (Bohool & Schmidt, 1974).

In a new approach to these problems, which makes no assumptions about the mechanism of the recognition step, the phenotype of mutant bacteria that are defective in the nodulation process, has been investigated. One such mutant, HS 111, of *Bradyrhizobium japonicum* strain USDA 110, has been shown to be unable to induce successful rapid nodulation (Halverson & Stacey, 1984). Pre-treatment of the mutant with an exudate obtained from soyabean roots for periods longer than 6 h reverses the defect in the effectiveness of the mutant, which then becomes nearly equivalent to that of good nodulating wild type bacteria. The existence of a factor has been demonstrated subsequently in root exudates; this is a lectin-like protein with N-acetyl-D-galactosamine/D-galactose-binding specificity active in the reversal of the nodulating defect in the mutant. Additionally, the lectin from soyabean seeds mimics the effect of the root factor and restores successful nodulation activity to the mutant (Halverson & Stacey, 1985). Galactose inhibits the improvement found in the nodulation characteristics of mutant HS 111 in keeping with the lectin-like properties of the root factor. Lectin–cell contact, therefore, appears to be obligatory for efficient nodulation with this defective mutant. Host specificity is clearly demonstrated by the finding that only the exudate from soyabean roots enhances the nodulation activity of the mutant and other seed lectins with the same, or similar specificity, i.e. lectins from roots of *Arachis hypogaea*, *Dolichos biflorus* or *Ricinus communis*, are ineffective (Halverson & Stacey, 1986b). Treatment of the wild type *Bradyrhizobium japonicum* USDA 110 of good nodulating activity with soyabean root exudates or the seed lectin increases the number of nodules formed on soyabean roots twofold at 6 weeks of age. It is also possible that lectin treatment may simply increase the number and proportion of bacterial cells capable of efficient nodulation in a population of symbiotic bacteria (Stacey *et al.*, 1986). Since treatment with the lectin for about 6 h is needed to correct the defect in the nodulation capacity of the mutant, it is unlikely that restoration of proper infectivity to the mutant is due to a simple absorption of the lectin on the bacterial cell surface. The lectin cannot restore full nodulating effectiveness if treatment of the mutant with the lectin is carried out in the presence of rifampin or chloramphenicol. Thus, exposure of the mutant to the lectin induces gene expression needed for efficient nodulation by the bacteria. After pre-treatment of the mutant with the lectin, the presence of four new proteins can be demonstrated by two-dimensional electrophoresis. Although these results appear to be very suggestive, the significance of changes in cellular protein content, and the relationship of the emergence of new proteins to the nodulation potential of the bacteria, still remain to be established (Stacey *et al.*, 1986).

Further supporting evidence for the involvement of lectins in the initial step of recognition and binding between root cells and host legume plants and appropriate rhizobial bacteria has been obtained from the studies of Kijne and his colleagues with *Pisum sativum* and *Rhizobium leguminosarum* symbiosis (Kijne *et al.*, 1986). Immunofluorescence microscopy has shown that lectins are localized almost exclusively on tips of young developing root hairs and epidermal cells just below the young root hairs; older root hairs, on the other hand, contain practically no lectin. Root sites containing lectins are relatively easy to infect while older root hairs without lectins are not susceptible to binding by *Rhizobium leguminosarum* (Diaz *et al.*, 1986). Also, a simple haptenic sugar, D-glucose, inhibits the attachment of the appropriate bacteria to root hair cells under special conditions (Kijne *et al.*, 1986). Finally, the maximum expression level of lectin mRNA in roots has been shown to be about 10 days after the start of germination by using a copy DNA of the pea seed lectin as a probe (Buffard *et al.*, 1988). The increased level of cell wall-associated lectin found by immunofluorescence on pea root hairs thus coincides with the maximum concentration of lectin mRNA (Diaz *et al.*, 1986).

One of the major difficulties encountered in attempting to compare results of the studies on the molecular mechanism of root hair attachment of bacterial cells obtained in different laboratories is that the main determinants of bacterial infectivity are the conditions of growth. Thus, nutrient limitation usually coincides with optimum attachment, also the type of nutrient limitation determines whether host lectins are involved in the binding of the bacteria (Smit, Kijne & Lugtenberg, 1986). Carbon limitation is known to induce a non-host-specific binding and neither the host plant lectin nor the *nod* genes appear to play any part in such an attachment (Smit *et al.*, 1986). In this carbon-limited rhizobial binding, the interaction between root hairs and the bacteria is through a calcium-dependent bacterial adhesion, followed by bacterial aggregation mediated by cellulose fibrils (Smit, Kijne & Lugtenberg, 1987). In yeast extract–mannitol media used for the growth of rhizobial bacteria in many laboratories, manganese ions are limiting bacterial growth and this growth limitation coincides with maximum attachment. More significantly, under conditions of manganese limitation, the binding and accumulation of rhizobial bacteria at root hair tips is accelerated by the presence of pea lectin molecules. Moreover, while carbon-limited bacteria are not infective, manganese-limited *Rhizobium leguminosarum* is. Thus, whether lectins localized on root hair tips can induce bacterial attachment and infection or not depends largely on the exact conditions of bacterial growth (Kijne *et al.*, 1988).

In a recently described elegant experiment, Diaz *et al.* (1989) have tested whether the participation of the root lectin in the early stages of the infection process, i.e. its recognition of, and the binding to, bacterial receptor molecules, is essential. In this experiment, the pea lectin gene encoding the D-glucose/mannose-specific pea lectin has been introduced into white clover roots by using *Agrobacterium rhizogenes* as a vector. As a result, the clover (*Trifolium repens*) roots have become susceptible to infection by *Rhizobium leguminosarum* bv. *viciae*. Although the 'hairy' roots have developed mostly pseudonodules on infection with the pea root-specific rhizobial strain, and in many cases the infection has aborted, the infection process has successfully proceeded beyond the root hair curling stage. This report is the first example of a host-specificity barrier being broken by genetic engineering of the host-plant (Diaz *et al.*, 1989).

Despite the impressive number of data gathered by the many research workers active in this important field which lends support to the essential nature of the lectin involvement in initiation of symbiotic reactions, there are still a number of questions which need to be answered satisfactorily before the lectin hypothesis can be generally accepted. An unequivocal demonstration of the presence of lectins and receptors on both plant root and bacterial cells at precisely the right time and location has yet to be obtained in all host–symbiont systems. Furthermore, a strict correlation between the specificity of the host lectin and its ability to recognize nodulating rhizobial bacteria which are specific for that host, needs to be clearly demonstrated. Also, the problematic existence of non-specific binding of non-homologous bacteria to plants needs to be satisfactorily resolved, perhaps along the lines demonstrated by Kijne and his colleagues with the dependence on growing conditions of the non-lectin-specific attachment of *Rhizobium leguminosarum* to pea roots, as discussed above (Kijne *et al.*, 1988). The failure of some lectin haptens to fully inhibit bacterial attachment must also be explained. The main questions to which answers still have to be found include some really fundamental ones. For example, are lectins necessary for all attachment or infection thread initiation? At exactly which step(s) are the lectins necessary? Thus, despite the great advances achieved in our understanding of the symbiotic relationship between plants and bacteria and the more-or-less general acceptance of lectin involvement at one stage or another in this process, there is still a clear need for further and more detailed studies.

Miscellaneous functions

In addition to the biological functions of lectins associated with their binding to endogenous or exogenous receptors as detailed in the previous sections, several other possible functions have also been envisaged for them. Unfortunately, most of these are rather hypothetical and not supported by adequate experimental evidence.

Thus, some experimental evidence appears to suggest that lectins may be components of the cell recognition system of higher plants. According to some current ideas, in the pollen–pistil, self-incompatibility system, which helps to maintain heterozygosity by favouring cross-pollination of plants, the success or failure of pollination depends on interactions between glycoproteins on the pollen surface and complementary products on the stigma or style (Heslop-Harrison, 1978). Some of the existing experimental evidence suggests that recognition between these two sets of proteins/glycoproteins may be mediated through lectins (Knox *et al.*, 1976). Most of the proteins and/or glycoconjugates obtained from pollens and stigma may function as β-lectins (Gleeson, Jermyn & Clarke, 1979) whose presence is demonstrated by their reactivity with artificial β-glycosyl (Yariw) antigens (Jermyn & Yeow, 1975). They occur widely in plant cell walls, sometimes bound to cell wall structures. Although it is debatable as to whether they fall within the general definition of lectins, β-lectins may function as non-specific adhesive substances and may provide physical support for pollen grains and the growing pollen tube (Knox *et al.*, 1976; Clarke, Anderson & Stone, 1979). Although some exogenously applied lectins, such as concanavalin A or PHA, appear to affect pollen germination and penetration (Southworth, 1975; Knox *et al.*, 1976), more thorough studies with homologous lectins will have to be carried out before the involvement of lectins in these processes can be accepted.

Apart from pollen–pistil interactions, cell wall lectins have been implicated in other physiologically important processes in plants. Thus, lectins have been found attached to plant cell walls non-covalently and it has been suggested that the main function of such lectins is to bind other components of the wall and to form bridges between underlying layers of the cell wall through cross-linked structures (Kauss & Glaser, 1974). It is envisaged that, as it may be possible to break and to reform the cross-links between different layers of the wall, such a process may be an important component in the mechanism of plant cell wall elongation. It is known that protons are produced during auxin-mediated plant cell wall elongation. As a result, the pH in the wall may be reduced sufficiently to provide the right conditions for the dissociation of the lectin from its ligands. Such weakening of the wall's rigidity may, indeed, be a prerequisite for the elongation. However, later experimental work has shown that D-galactose, the specific hapten-sugar inhibitor of the cell wall lectin from mung bean (*Vigna radiata*), does not dissociate or solubilize the lectin from cell wall preparations of this plant (Haass *et al.*, 1981). Moreover, the hapten sugar does not inhibit the binding of the isolated lectin to cell wall preparations.

Similar ideas have also been proposed for the physiological function of the leaf-stem CRM lectin in *Dolichos biflorus*. Thus, the leaf-stem lectin of this plant is, at least in part, associated with the cell wall (Talbot & Etzler, 1978; Etzler *et al.*, 1984) and its amount increases when the plant is infected or wounded (Etzler, 1986). Accordingly, this leaf-stem lectin may be involved in cell wall metabolism of the plant. As heat shock appears to enhance the synthesis of another lectin-related protein (DB57) in *Dolichos biflorus* cell suspension cultures (Spadoro-Tank & Etzler, 1988) or to promote the synthesis of PHA in the endoplasmic reticulum in *Phaseolus vulgaris* (Chrispeels & Greenwood, 1987), lectins may play a role in the defence of plants against adverse environmental and/or biological agents.

The involvement of lectins in cell wall metabolism of sugar beet roots has also been suggested since two lectins have been isolated from sugar beet root cell wall fractions by extraction with neutral detergents and EDTA. In addition, it has been demonstrated that an endogenous receptor is present tightly bound to microsomal membranes. It is possible that the sugar beet root cell wall lectins may participate in an intracellular sucrose-transporting system. However, such an idea, at present, is mainly conjectural (Aleksidze *et al.*, 1983; Aleksidze & Vyskrebentseva, 1986).

As most plant lectins are strongly mitogenic for peripheral mammalian lymphocytes, it is possible that seed lectins may be mitotic stimulants for plant cells. However, the experimental evidence appears to be against such suggestions. For example, the soyabean lectin is a very weak mitotic stimulant for callus cells and is without any effect for the explants of soyabean roots (for refs see Pusztai *et al.*, 1983*a*; Etzler, 1986). Moreover, the results of similar experiments with lectins and protoplasts have been almost totally negative.

Before the more recent definitions for lectins had been universally accepted, it was proposed that lectins may be similar to some glycosidases and, vice versa, that some glycosidases may function as lectins. In fact, there are impressive and extensive immunochemical and molecular similarities between some lectins and, for example, a number of plant α-galactosidases capable of agglutinating red cells. Accordingly, it is possible that an evolutionary relationship between glycosidases and lectins exists (Hankins, Kindinger & Shannon, 1979, 1980*a,b*). However, the agglutination caused by glycosidases is usually only temporary and the clot disperses after the completion of the glycosidase enzyme action with the removal of the surface sugar. Thus, glycosidase enzymes are not true lectins and as such are, by definition, specifically excluded from the lectin class of carbohydrate-binding proteins.

Some other plant glycosidase enzymes may justifiably be included with true lectins. For example, the D-galactose-specific cell wall lectin from *Vigna radiata*, which also functions as an α-galactosidase enzyme, appears to be distinctly different from the α-galactosidases described by Shannon and his colleagues. The cell wall lectin is a tetrameric glycoprotein, $M_r = 170,000$, composed of two distinct subunits, one of which is a lectin, while the other has activity for hydrolysing α-galactosides (Haass *et al.*, 1981). The physiological significance in the plant of a protein with such dual biological activity is unknown.

A potentially more interesting example of mixed-function, lectin-enzyme proteins has been isolated from the seeds of broad bean, *Vicia faba*, and its properties described (Dey, Naik & Pridham, 1982*a*; Dey, Pridham & Sumar, 1982*b*; Dey, Naik & Pridham, 1986). This protein can, apparently, function both as an α-galactosidase enzyme and as a D-mannose/glucose-specific lectin. As the two biological activities appear to be independent of each other, the conclusion is inescapable that the protein molecule must have two different carbohydrate recognition sites, one for α-linked galactosyl residues and another for D-mannose/glucose residues (Dey *et al.*, 1986). Thus, this particular lectin-enzyme hybrid protein may have to be seriously considered as a member of the family of plant lectins.

Finally, there is a large body of experimental evidence to suggest that some lectins from seeds (Pusztai *et al.*, 1983*a*; Etzler, 1986), bark tissues (Greenwood *et al.*, 1986; Nsimba-Lubaki & Peumans, 1986) and rhizomes (Peumans *et al.*, 1985) are localized in protein bodies of the respective tissues of plants. By definition, these lectins may be regarded as storage proteins. The amounts of such lectins, for example, from the bark, usually display definite seasonal fluctuations since they accumulate in the autumn and are depleted during the spring. Furthermore, seed lectins have for long been regarded as specialized forms of storage proteins, which may occasionally have other functions. Indeed, lectins are included in the general group of seed storage proteins on account of their usually appreciable occurrence in seed protein bodies, the timing of their biosynthesis during seed development and similarities between their behaviour and that of storage proteins during germination (Pusztai *et al.*, 1983*a*).

5

Effects on blood cells

The realization made almost simultaneously by Boyd (Boyd & Reguera, 1949) and Renkonnen (1948) that some plant lectins are blood group-specific has done more than any other to turn the studies of lectins from an interesting, though somewhat esoteric field of study into a major and multidisciplinary science. Thus, Nowell discovered in 1960 that preparations of a lectin from the seeds of *Phaseolus vulgaris* (PHA) have strong mitogenic activity on peripheral lymphocytes (Nowell, 1960). Not long after this, Aub and his colleagues recognized by a chance observation that some transformed cells have greater affinity for lectins than the cell lines from which they originate (Aub, Tieslau & Lankester, 1963). Obviously, both these important effects are the result of specific reactions between cell surface carbohydrate structures and the appropriate lectins. This initial specific recognition of and binding to cell surface receptors by lectins occurs both *in vivo* and *in vitro* and has wide-ranging and dramatic effects on cells.

These two major discoveries led to an upsurge of research activity by scientists from all fields of biological sciences, particularly, cell biologists interested in immunology and cancer research. As the composition and the physical state of the cell surface is known to change throughout the life-cycle of the cell and to alter radically during cancerous or pre-cancerous transformation, it is not surprising to find that lectins, which can specifically recognize such surface receptors, are exploited as well-adaptable and excellent tools for monitoring and studying these changes.

Agglutination

Agglutination is one of the most characteristic and important reactions of those plant lectins which have at least two carbohydrate-binding sites. Indeed, according to the earlier and more restricted definition of lectins, of the great numbers of carbohydrate-binding proteins, only those which agglutinate cells were called lectins (Goldstein *et al.*, 1980). Even more to the point, it is widely known that Stillmark discovered lectins by observing their agglutination activity (Stillmark, 1888).

Although studies concerned with lectin-induced cell agglutination have had a long history, not all important parameters controlling this complex reaction are understood fully. For example, although it is known that binding of lectins to the surface of cells is a necessary prerequisite for agglutination, the extent of lectin-binding does not always correlate fully with the actual proportion of cells agglutinated. In most instances, the extent of agglutination is dependent on the specificity, valency, size of the lectins, and whether they can interact with both glycolipids and glycoproteins. Naturally, agglutination is also dependent on the cell, its surface properties and metabolic state. The extent of agglutination is particularly influenced by the nature, number, distribution, exposure and mobility of the surface receptors on the cell, whilst additional factors may include the deformability, fluidity and surface charges of the cell membrane (Nicolson, 1974). By the manipulation of some or most of these factors, or by selecting the most appropriate physical conditions, the selectivity of agglutination of individual cell types can be increased to such an extent that particular cells can be separated and isolated from complex mixtures of different cell types. In fact, appropriate and careful application of this method provides us with one of the simplest cell fractionation techniques. However, although the extent of the agglutination may serve as a measure of changes occurring in the surface properties of cells, the actual nature of the change is difficult to determine from agglutination measurements only because of the complexity of the factors controlling the reaction.

It is not always realized that measurements of the extent of agglutination are not always carried out under equilibrium conditions and, hence, to obtain valid comparisons, the assay conditions, concentration of cells, time, temperature and pH, have to be rigorously standardized. The agglutination rate is also known to be dependent on the concentration of the lectin and the density of the cell suspension used. In fact, the relative proportions of the two partners of the agglutination reaction are similarly important. Moreover, the temperature of the reaction is vital because the mobility of the membrane receptors is temperature-dependent, and also because the rate of internalization of the lectins by cells are, to a large extent, determined by the temperature of the reaction.

In most instances, the extent of agglutination of cells, such as the erythrocytes, can be increased by various pre-treatments with the appropriate enzymes. In most agglutination reactions, the main site of interaction is between the surface glycolipids of cells and the lectin. If, for example, peptide chains protruding from the lipid bilayer of surface membranes are removed by proteolysis (e.g. by trypsin, papain, pronase, etc), the closer contact possible between the lectin and the cell surface usually increases the extent of agglutination. One of the best examples for demonstrating this is to be found in the known behaviour of the *Phaseolus vulgaris*, Pinto III-lectin which, in contrast to most tetramer PHA isolectins, is a dimer and therefore is a poor agglutinin for most untreated red cells. However, its agglutination potency improves considerably with pronase-treated red cells (Pusztai *et al.*, 1981*a*). Similarly, extracts from other legume seeds which contain lectins of low haemagglutination activity, such as, for example, *Vigna unguiculata*, *Cicer arietanum*, etc, usually show much higher activity with pronase-treated rat cells (Grant *et al.*, 1983). The removal of negatively charged sialic acid residues by treatment of red cells with neuraminidase may also increase cell agglutinability as a result of the reduction of electrostatic repulsion between the cells and lectins. In addition, removal of terminal sialic acid residues may also uncover new potential receptors in the cell membrane, leading to more extensive binding and agglutination. For example, peanut lectin reacts with human erythrocytes only after the sialic acid is removed from the red cells by neuraminidase treatment (Lis & Sharon, 1981).

Moderate aggregation of the lectin may increase the extent of agglutination by generating more points of contact between the reacting partners, i.e. the lectin and the cells. Thus, for example, the soyabean lectin is a more potent agglutinin than its protomer when slightly polymerized with glutaraldehyde (Tunis, Lis & Sharon, 1979). In contrast, the dimer succinyl–concanavalin A derivative is about 500 times less active in haemagglutination tests than the native tetravalent concanavalin A (Wang & Edelman, 1978). However, it is by no means certain that the reduction in the haemagglutination activity of the dimer succinyl–concanavalin A is solely due to reduction of the size of the lectin. Quite possibly, the introduction of a great number of negatively charged succinyl groups may lead to increased repulsion between the reaction partners and, therefore, a reduction in the extent of haemagglutination. This point is further supported by the finding that the sulphomethylamide derivative of concanavalin A, which is also a dimer but has a similar overall charge to the native tetramer lectin, is only about six to seven times less active in haemagglutination tests (Saito *et al.*, 1983; Ishii *et al.*,

1984). Rather interestingly, even the monovalent monomer form of concanavalin A prepared by photochemical alkylation or H_2O_2/dioxane oxidation shows well-demonstrated haemagglutination activity when tested with guinea pig red cells (Ishii *et al.*, 1984). Although this derivative is about a hundred times less active as a haemagglutinin than the native tetramer concanavalin A, its definite capacity to agglutinate red cells suggests that the monovalent lectin may also possess hydrophobic sites capable of interacting with similar sites on the surface of erythrocytes in addition to its single carbohydrate-binding site. In this sense, this concanavalin A derivative must have at least two valencies of different nature for binding ligands (Ishii *et al.*, 1984).

Changes in surface carbohydrate structures on cells usually lead to differences in their agglutinability, regardless of whether they have been induced by transformation from normal into malignant (Aub *et al.*, 1963), by maturation from embryonic into adult or simply by their mitotic stimulation (Rapin & Burger, 1974). In contrast, modifications in the carbohydrate side-chains of glycoprotein lectins have little effect on their agglutinating potency (Lis & Sharon, 1981) and changes in cell surface carbohydrates are not necessarily associated with increased agglutination. For example, it is not generally valid that all transformed cells are agglutinated more extensively than their normal counterparts (Ukena *et al.*, 1976; Reisner, Sharon & Haran-Ghera, 1980) although the agglutinability of Ehrlich ascites cells by concanavalin A, or that of lymphocytes by PHA (Chatterjee, Guha & Chattopadhyay, 1986), increases with the development of the tumour in mice (Chakraborty, Bose & Chowdhury, 1987).

Finally, not just animal cells but those from higher plants, microorganisms, fungi, algae and other species can also be agglutinated by lectins. Sumner and Howell showed as early as 1936 that concanavalin A agglutinates bacteria from the genera *Mycobacterium* and *Actinomyces* (Sumner & Howell, 1936) and, since then, there has been considerable progress in our understanding of lectin–microbe interactions (Pistole, 1981). With this advance, it is not surprising that lectins have been proposed as diagnostic tools in microbiological studies (Doyle & Keller, 1984).

There is experimental evidence to indicate that subcellular organelles can be agglutinated by lectins. Even isolated vacuoles from plant cells, such as the red beet (*Beta vulgaris*) are agglutinated by a number of lectins (Salyajev & Kuzevanov, 1984), which suggests that there are lectin receptors present on the tonoplast, the vacuolar membrane of this organelle.

Mitogenic stimulation of lymphocytes

One of the most momentous discoveries made with lectins was Nowell's observation that kidney bean lectins, or very crude PHA preparations, even seed extracts, are powerful stimulants for quiescent peripheral lymphocytes and induce blast formation leading to mitosis (Nowell, 1960). It is now recognized that a great number of complex intracellular changes take place as a result of lectin-binding to lymphocytes, and to other cells, and that all these follow on from the initial binding of the lectins and other mitogenic agents to the membrane of blood cells. It is suspected that, under appropriate and favourable conditions, most plant lectins are mitogenic (Lis & Sharon, 1986). It has been shown that even the asialoglycoprotein receptor (lectin) from rabbit liver hepatocyte membrane is mitogenic for neuraminidase-treated lymphocytes (Novogrodsky & Ashwell, 1977) as are some lectin binders from plants (Gebauer *et al.*, 1982).

The mitogenic activity of lectins, similarly to that found for their haemagglutination reactions, can be inhibited specifically by the appropriate haptenic carbohydrates. However, it is clear that the function of lectins in mitosis is a great deal more complex than that in cell agglutination. Some lectins, for example, wheatgerm agglutinin, may antagonize the mitotic activation of lymphocytes induced by a number of stimuli at appropriate lectin concentrations (Barrett *et al.*, 1983). Tomato lectin is anti-mitogenic for chicken but not for mouse lymphocytes (Nachbar *et al.*, 1980). *Datura* lectin, although mitogenic on its own, is also anti-mitogenic under certain conditions and may antagonize blast formation induced by a purified protein derivative of tuberculin. Furthermore, the mitogenicity of this lectin for human lymphocytes is enormously enhanced in the presence of the phorbol ester, TPA. In contrast, TPA has no such synergistic effect on the mitogenicity of the structurally related potato or tomato lectins (McCurrach & Kilpatrick, 1988). Tomato lectin is not only non-mitogenic for human lymphocytes by suppressing spontaneous DNA synthesis in these cells, but it is also anti-mitogenic for lymphocytes stimulated by recall antigens or allogeneic cells *in vitro* (Kilpatrick, Graham & Urbaniak, 1986). Tomato lectin is able to bind several membrane glycoproteins from lymphocytes and can agglutinate both B and T lymphocytes. However, the lectin does not bind to major histocompatibility antigens (HLA), quite unlike the known mitogenic lectins, which avidly bind to either class I or class II HLA antigens (Kimura & Ersson, 1981).

A further complication is that some lectins may be mitogenic at low concentrations, while anti-mitogenic at higher concentrations. Furthermore, lectins such as wheatgerm agglutinin usually require the presence of accessory cells for mitogenic activity. Thus, the mitogenicity of these lectins is dependent on the serum used to supplement the culture medium (Kilpatrick & McCurrach, 1987). Wheatgerm agglutinin may be considered as an 'incomplete' mitogen as it produces a quantitatively similar effect to that obtained with the classical mitogen, PHA, only in the presence of the tumour promoter, 12-*o*-tetradecanoyl-13-acetate (TPA). Interleukin-2 has a somewhat similar synergistic effect on the mitogenicity of wheatgerm agglutinin, but it causes only a partial improvement in T cell response to the lectin (Kilpatrick & McCurrach, 1987).

Initially, it was believed that lectins are mitogenic for T-lymphocytes only. However, closer investigations have revealed that some lectins can stimulate both T and B lymphocytes. For example, pokeweed mitogen activates both types of lymphocytes (Basham & Waxdal, 1975). Similar observations have been made with lentil lectin (Miller, 1983).

In addition to lectins, several other agents are mitogenic. The most important of these are those antigens which specifically induce proliferation in individual clones of lymphocytes. In contrast, lectins do not activate lymphocytes in this antigen-like, selective manner. Lectins are polyclonal signals. They bind to carbohydrate receptors on the surface of all those lymphocytes which contain the same sugar structures appropriate for the specificity of the mitogenic lectin. It is frequently found that, on stimulation with some lectins under appropriate conditions (Hume & Weidemann, 1980) over 80% of the total lymphocyte population may be induced to mitosis. Thus, lectins have a clear advantage over specific antigens as mitogenic stimulants and they are the best polyclonal stimulants for studies into the reaction mechanism of the mitogenic transformation of lymphocytes.

Similar to that happening in lectin-induced agglutination reactions, the first step in the mitogenic stimulation of lymphocytes is the binding of the lectins to surface glycoconjugates of the cells. The binding occurs rapidly and is followed by patching and capping. In some instances, the whole surface glycoconjugate–lectin complex may then be endocytosed by the cell. Such patching/capping also occurs in platelet membranes after stimulation with concanavalin A (Kakaiya, Kiraly & Cable, 1988). Thus, these membrane reactions are probably a part of the general early response of cells to exposure to lectins.

There is good experimental evidence to indicate that both antigens and lectins bind to the same receptor on T lymphocytes (Chilson, Boylston & Crumpton, 1984). This initial step of binding is the trigger for the entire mitogenic process. Accordingly, it is expected that mitogenic and non-mitogenic lectins ought to bind to different lymphocyte membrane glycoconjugates if they were to send different signals to the inside of the cells. Although initial attempts to demonstrate this have been unsuccessful (Skoog, Nilson & Weber, 1980), the recent finding, already mentioned above, that mitogenic lectins bind to components of the major histocompatibility antigens (Kimura & Ersson, 1981), while non-mitogenic lectins generally do not (Kilpatrick *et al.*, 1986), indicates that, indeed, such differences in binding patterns do exist. Differences in binding between mitogenic and non-mitogenic lectins have also been found in similar studies with neuraminidase-treated lymphocyte membrane glycoproteins (Dillner-Centerlind *et al.*, 1980). Thus, the initial trigger-signal concept is now experimentally established and is compatible with evidence based on the glycoprotein composition and topology of lymphocyte membranes. It may be envisaged that, as both mitogenic and non-mitogenic lectins bind to lymphocytes, in addition to the presence of receptor clusters capable of triggering stimulatory responses, other regions in the lymphocyte membranes may contain inhibitory domains.

After binding of the mitogenic lectin to the lymphocyte membrane, the first effect of the stimulatory signal is an increase in membrane permeability and phospholipid metabolism. Although the rates of phospholipid methylation reactions are increased, the concurrent degradation rate of methylated phospholipids is also accelerated (Toyoshima *et al.*, 1982). Several signal molecules, such as arachidonic acid and its derivatives including prostaglandins and thromboxanes, are produced through the action of phospholipase A; the role of such signal molecules in mitogenic transformation is now generally accepted.

As a first step in T cell proliferation, mitogenic monoclonal antibodies against the T3 molecule on the surface membranes of T lymphocytes or mitogenic lectins induce activation of ornithine decarboxylase (ODC) within 5 min without affecting protein synthesis in the cell. Non-mitogenic ligands do not induce such ODC activation. The rapid activation of ODC requires energy and an intact cytoskeleton, and is associated with the mitogen-stimulated breakdown of phosphoinositides (Mustelin *et al.*, 1986). The mitogen-induced activation of ODC is abolished in T cells, which are first selectively depleted of guanine nucleotides by treatment with mycophenolic acid. The involvement of G protein(s) in the transduction of the proliferation signal is also suggested by the restoration of ODC induction if guanine is added to the depleted T cells just before the mitogenic stimulus. Non-hydrolysable analogues, such as guanosine-5'-(3-*o*-thio) triphosphate, GTPγS, are also effective. The precise identity of the G protein involved in ODC induction is not known but some recent results suggest that *ras*-proto-oncogene encoded proteins (p21), which have been implicated in phosphatidylinositol turnover and cell proliferation, may also be responsible for ODC induction (Mustelin, 1987).

The great importance of G proteins located in lymphocyte membranes for the mitogenic signal transduction process as coupling agents between membrane receptors and intracellular effectors, is now generally recognized. Members of this family of guanine nucleotide (GTP)-binding proteins are all involved in receptor-mediated transmembrane signalling processes. Mitogenic signals, such as concanavalin A, may be coupled to adenylate cyclase through G_s (stimulatory G protein) and G_i (inhibitory G protein) proteins. Another transducer G protein is involved in phosphoinositide metabolism acting as a coupler between membrane receptors and phosphoinositide-specific phospholipase C (PI–PLC). Accordingly, stimulation with concanavalin A leads to a rapid redistribution of a part of the GTP-binding activity from membranes to the cytosol in murine thymocytes. The extent of stimulation of cytosolic PI–PLC activity obtained by supplementing the cells with GTPγS is appreciably higher in concanavalin A-stimulated thymocytes than in the non-stimulated ones, and both these changes are dependent on the concentration of concanavalin A used for the stimulation. It is possible that the translocation of the G protein from the membrane to the cytosol is also responsible for the activation of phospholipase C enzyme. A protein fraction obtained from stimulated thymocyte membranes, with subunit M_r values between 23,000 and 28,000, has been shown to be an effective activator of cytosolic PI–PLC, suggesting that this GTP-binding subunit may be a subunit of the G protein, which is translocated from the membrane to the cytosol and is responsible for phospholipase C activation upon stimulation of the thymocytes with concanavalin A (Wang, Toyoshima & Osawa, 1988).

More evidence for the existence of direct interactions between concanavalin A receptors and G proteins has been obtained recently. Thus, the inhibitory GTP-binding protein, G_i, binds to concanavalin-A Sepharose-4B columns. However, in membrane extracts prepared from thymocytes stimulated with concanavalin A, no such affinity-absorbed G_i protein exists, suggesting that, on stimulation with concanavalin A, this G_i protein dissociates from the membrane. Moreover, this G protein abolishes the inhibitory effect of concanavalin A on the prostaglandin E_1- and isoproterenol-induced increases in cellular cAMP. Accordingly, concanavalin A receptor(s)-associated G_i may be the G protein responsible for the inhibitory regulation of cellular cyclic AMP levels. These results give a clear and more direct indication that the G_i protein, together with the 23,000 to 28,000 molecular weight G protein subunit, interacts with the membrane receptor(s) of the original mitogenic signal, concanavalin A (Wang *et al.*, 1989).

Further understanding of the early events of signal transduction after lectin-binding to cell surface receptors has come from analyses of changes in membrane phosphoinositide turnover. Hydrolysis of phosphatidyl inositol 4,5-diphosphate (PIP_2) leads to the generation of second messengers, diacylglycerol (DAG) and inositol 1,4,5-triphosphate (IP_3) (Nishizuka, 1984; Berridge, 1984, 1987; Whitfield *et al.*, 1987). The hydrolysis of PIP_2 proceeds normally, leading to increase in cytosolic Ca^{2+} concentration even after the binding of incomplete mitogens, such as wheatgerm agglutinin. However, for the adequate expression of the second signal, the presence of submitogenic amounts of TPA is also necessary. Wheatgerm agglutinin becomes a complete mitogen only when low doses of TPA are present (Clevers *et al.*, 1986). The enzyme, phosphoinositide phosphodiesterase, or briefly, phospholipase C, occurs both in cell membranes (Cockroft, 1987) and the cytosol (Majerus *et al.*, 1986) and the indications are that it is mainly the membrane-bound enzyme which is responsible for the generation of the phosphoinositide-derived intracellular second messengers. The overriding importance of phospholipase C is shown by the observation that treatment of human lymphocytes with monospecific anti-phospholipase C antibodies causes a dose-dependent inhibition, up to 100%, of the lectin-induced proliferation (Moraru *et al.*, 1987). Such a strong correlation between phosphoinositide hydrolysis and lectin-induced mitogenesis underlines the importance of the intracellular second messengers generated in the process. Thus, production of IP_3 first triggers the release of Ca^{2+} from the endoplasmic reticulum (Berridge *et al.*, 1985; Berridge, 1987), which is followed by the opening of the plasma membrane calcium channels (Whitfield *et al.*, 1987). On stimulation with PHA, the second product of phosphoinositide hydrolysis, DAG, increases the activity of the phospholipid-dependent protein kinase C both in the cytosol and the membrane fraction in the presence of the released Ca^{2+} (Mire, Wickremasinghe & Hoffbrand, 1986*a*). Although the increased free intracellular calcium concentration resulting from phosphatidyl inositol hydrolysis and the generation of second messengers is thought to make a major contribution to later cellular responses of the proliferative process, such as, for example, lymphokine gene expression, short-term increases in calcium concentration do not always lead to complete signal transmission and full cellular response. For example, in T cell somatic mutants with deficient receptor function, a monoclonal antibody produced against the receptor complex failed to promote late events of the mitogenic process despite a very substantial and early mobilization of intracellular Ca^{2+} (Goldsmith & Weiss, 1988).

On stimulation with PHA, the increase in the activity of the membrane-bound protein kinase C enzyme is different from that found with phorbol ester (TPA) activation (Castagna *et al.*, 1982). TPA and, to a certain extent, interleukin-2 or interleukin-3 (Farrar & Anderson, 1985; Farrar, Thomas & Anderson, 1985) stimulate the activity of the membrane-bound protein kinase C by promoting its translocation from the cytosol to the particulate fraction, with a consequent fall in the enzyme concentration in the cytosol. In contrast, in human peripheral blood lymphocytes after stimulation with PHA the total cellular phosphokinase C activity increases (Mire *et al.*, 1986*b*) resulting in a more sustained and elevated membrane-bound enzyme response than that obtained with phorbol esters or interleukins.

This sustained effect of the lectin on the activity of protein kinase C may not be generally valid for all cells. For example, in primary bovine lymphocytes, both concanavalin A and phorbol esters induce a very quick (5 min) three- to fourfold increase in the activity of the particulate protein kinase C. However, this is at the expense of a 30–50% decrease in the concentration of this enzyme in the cytosol (Grove & Mastro, 1987). Moreover, in rat glioma C_6 cells, concanavalin A actually prevents the TPA-induced redistribution of protein kinase C from the cytosol into the particulate fraction (Patel & Kassis, 1987). On the other hand, in thymoma cells from a patient with myasthenia gravis, the translocation of protein kinase C is different when the stimulant is concanavalin A or the phorbol ester. Thus, although protein kinase C activity increases in the cytosol on concanavalin A stimulation, it is reduced in the membrane fraction. In contrast, treatment with TPA decreases the activity of the membrane-bound enzyme, while increasing it in the cytosol (Ishizuka *et al.*, 1988). The complexity of this reaction pathway is shown by the demonstrably anti-proliferative effect of PHA on human leukaemic cells. Although stimulation of these cells with PHA increases the activity of cytosolic protein kinase C, this is not the result of its translocation from the particulate fraction. In addition, there is a good correlation between this initial protein kinase C-induction by PHA and its anti-proliferative effects (Borrebaeck & Schon, 1987; Borrebaeck, Bristulf & Jergil, 1987). The precise mechanism of the increase in the activity of the membrane-bound functional protein kinase C in stimulated lymphocytes is still not clear and the fine details of this process need not necessarily be the same for all different cell types. It is possible that the main difference between stimulation with complete mitogens such as concanavalin A and incomplete mitogens, such as the phorbol esters (TPA), is in their different effects on the activation of protein kinases. Although both mitogens produce an initial and quick response of protein kinase C translocation from the cytosol to the particulate fraction, which is necessary for protein phosphorylation in the next step of the mitogenic response, TPA has no effect on cytosolic cAMP-dependent protein kinases. On the other hand, the activity of protein kinase II is appreciably increased on mitogenic stimulation with concanavalin A. Accordingly, the activation of both types of protein kinases appears to be necessary for successful mitogenicity (Grove & Mastro, 1987).

The importance of other signal pathways, such as the cAMP-sensitive signalling system in which Ca^{2+} and protein kinase C are not directly involved has been revealed recently. In addition to increasing intracellular Ca^{2+} concentration and phosphoinositide turnover, concanavalin A is also responsible for the rapid increase in cytosolic pH and Na^+ ion concentration. Apparently, the stimulation of the Na^+/H^+ antiport by concanavalin A can proceed unimpeded in the absence of adequate intracellular calcium and/or even when the activity of protein kinase C is blocked (Grinstein *et al.*, 1987). In fact, the existence of such a third signalling pathway has also been suggested by the observation that lectins can induce proliferation in TPA-treated cells by a mechanism independent of Ca^{2+} (Gelfand *et al.*, 1985). Moreover, a significant potentiation of the proliferative response to exogenous DAG and Ca^{2+} ionophore can be obtained by further addition of lectins (Kaibuchi, Takai & Nishizuka, 1985). It is possible that this third signalling pathway may function by the activation of the tyrosine kinase of the lectin receptor. Unfortunately, how the third signalling pathway is related to Na^+/H^+ exchange mechanism or, indeed, to proliferation, still needs to be unravelled.

The complexity of the relationship between the activation of the Na^+/H^+ antiport system and mitosis is clearly indicated by the findings that, although concanavalin A stimulates the activation of the antiporter and simultaneously induces the induction of ornithine decarboxylase (ODC) and DNA synthesis, both of which are necessary for proliferation, the activation and/or inhibition of the three processes can be separated under different experimental conditions. Thus amiloride, a specific inhibitor of the antiporter does not inhibit the induction of ODC and DNA synthesis in the presence of bicarbonate in the medium. In contrast, although the internal pH is increased by concanavalin A in the absence of bicarbonate, amiloride inhibits both the antiporter and DNA synthesis, but not the induction of ODC. These results suggest that the activation of the

Na^+/H^+ antiport system is not a trigger for cell proliferation, but that it is probably needed for the maintenance of the internal pH optimum, especially in bicarbonate-free media (Kakinuma *et al.*, 1987).

The importance of the induction of enzymes involved in polyamine metabolism during mitosis has been underlined. It appears that, of the three enzymes which regulate polyamine levels in cells, ornithine decarboxylase (ODC), *S*-adenosylmethionine decarboxylase (SAMDC) and spermidine/spermine N^1-acetyltransferase (SAT), at least two, ODC and SAT, are induced through the action of protein kinase C generated by transmembrane signalling reactions. It has been suggested that regulation of DNA synthesis during mitosis is, at least in part, achieved by appropriate changes in the concentration of intracellular polyamines. Accordingly, one of the post-receptor signalling pathways linking the mitogenic signal to DNA synthesis and proliferation may be through the activation of protein kinase C, coupled to the induction of polyamine metabolic enzymes (Otani *et al.*, 1986; Matsui-Yuasa, Otani & Morisawa, 1987).

Phosphoinositide metabolism and changes in intracellular calcium concentrations and all related responses to the activation by the polyclonal mitogen, concanavalin A, are very similar in both normal murine thymocytes and their transformed counterparts, such as S49-T-lymphoma cells (Taylor, Hesketh & Metcalf, 1988). The inhibition by cyclic AMP of concanavalin A-induced increase of Ca^{2+} concentration and its dependence on cAMP concentration and extracellular Ca^{2+} levels is similar in both normal thymocytes and S49 cells. Moreover, when re-stimulated with concanavalin A after the removal of the initial stimulus, the secondary lymphocytes produced from activated primary lymphocytes responded with a highly accelerated Ca^{2+} uptake and increased DNA synthesis (Wolff, 1987). Accordingly, it appears that phospho-inositide hydrolysis and the generation of second messengers are obligatory steps in the mitogenic stimulation of peripheral lymphocytes by mitogenic lectins.

The lectin-induced membrane reactions lead not only to the generation of second messengers but also to other profound changes in the activated lymphocytes. Mitogenic stimulation induces the synthesis of specific proteins not found in resting T lymphocytes. These include a number of immunoregulatory lymphokines, such as interleukin-2, (IL-2). Other factors which are capable of controlling the growth process, such as the receptors for interleukin-2, (IL-2R) (Lipkowitz *et al.*, 1984; Cantrell & Smith, 1983) and for transferrin (Neckers & Cossman, 1983) are also produced. After stimulation with PHA, the resting lymphocytes of G_0 phase progress to G_1 phase of the cell cycle. At this point, the stimulated T cells begin to synthesize both IL-2 and its receptor, IL-2R. Those activated lymphocytes which produce the receptor, IL-2R, bind the simultaneously synthesized IL-2, which, in turn, leads to the production of transferrin receptors (Neckers & Cossman, 1983; Kronke *et al.*, 1985).

Although resting lymphocytes contain intracellular receptors for serotransferrin (Weiel & Hamilton, 1984), they do not express either surface or intracellular receptors for lactotransferrin. In contrast, human peripheral blood lymphocytes express both surface and intracellular lactotransferrin receptors after stimulation with PHA. The process is time-dependent and maximum receptor concentration is usually attained by about two days after the original stimulation. This is not achieved by redistribution or modification of existing receptors. Apparently, resting lymphocytes possess a gene for lactotransferrin receptor, although this is only expressed following a mitogenic stimulus. Indeed, one of the effects of the PHA signal is the induction of receptor biosynthesis as a part of the mitogenic process. Lactotransferrin receptors which are fully active have been purified from Triton X-100 extracts of PHA-stimulated lymphocytes by anti-ligand-affinity chromatography. The binding of the receptor to lactotransferrin is reversible and concentration and pH dependent (Mazurier *et al.*, 1989).

The binding of transferrin to T cells is an essential part of the mitogenic process and it is also clear that human lactotransferrin possesses growth-stimulating activity for human lymphocytes after PHA treatment in a way similar to that previously described for human serotransferrin (Brock, 1981). Stimulated T cells advance from the late G_1 into the S phase of the cell cycle in the late phase of mitosis as a result of the binding of transferrin by its receptor (Neckers & Cossman, 1983).

There are a number of important differences in the expression of IL-2, IL-2R and transferrin-receptor genes in proliferating human T lymphocytes dependent upon whether the mitogenic activation has been induced by PHA or phorbol esters plus calcium ionophore. In comparison with that occurring with PHA, the expression of both IL-2 and transferrin-receptor genes are accelerated when T cells are stimulated by the phorbol ester in combination with the calcium ionophore. In addition, the combined effects of the phorbol ester plus ionophore also trigger an IL-2-independent pathway for the induction of transferrin-receptor mRNA (Kumagai *et al.*, 1988).

The effects of wheatgerm agglutinin on IL-2 production and expression of its receptor, IL-2R, have been fully explored. The responsiveness of stimulated T lymphocytes to the IL-2 produced during stimulation with wheatgerm agglutinin is particularly interesting as it is known to have only slight activity for human T lymphocytes on its own. Moreover, although it inhibits the PHA-induced mitogenic response of lymphocytes, wheatgerm agglutinin is as good an inducer of IL-2 mRNA production as PHA and both lectins have similar effects on cytoplasmic-free Ca^{2+} concentration. Additionally, about 20% of the wheatgerm-stimulated lymphocytes express IL-2 receptors although PHA-induced IL-2R expression is not blocked by wheatgerm agglutinin. If wheatgerm agglutinin is removed from lymphocyte cultures, the cells start to proliferate after the addition of exogenous IL-2, whilst the IL-2-dependent proliferation of PHA-blasts can be blocked by the addition of wheatgerm agglutinin. These results taken together with the known binding of wheatgerm agglutinin to lymphocyte IL-2 receptors (Read *et al.*, 1985), indicate that the lectin inhibits T cell proliferation by inhibiting their responsiveness to IL-2 and not by interfering with the production of IL-2 or expression of IL-2R (Kawakami, Yamamoto & Onoue, 1988).

The requirements for the induction of IL-2R expression in lymphocytes during *in vitro* stimulation are reasonably well understood, but the exploration of the roles of IL-2 and its receptor, IL-2R, during the *in vivo* activation of lymphocytes is more recent. Thus, a model has been developed for the rapid local stimulation of popliteal lymph nodes by subcutaneous injection of concanavalin A in mouse footpads. Both major subsets of T cells ($Lyt-2^+$ and $L3T4^+$) and B cells in the popliteal lymph nodes express both 1L-2R and transferrin receptor and at peak response, 9–15 h after the injection of concanavalin A, about 70% of all these cells express the receptors. The IL-2R expression is closely associated with both spontaneous and IL-2-driven proliferation (Black *et al.*, 1988) and it is quite likely that the mechanisms controlling the induction of mitogen responsiveness are similar both *in vivo* and *in vitro*. However, the response is rapidly diminished *in vivo* with the reduction of the initial immunostimulatory signal due to the decrease in the concentration of the mitogen, concanavalin A, in the lymph nodes. Additionally, it is also known that concanavalin A *in vivo* appears to induce appreciable numbers of T suppressor cells whose function may involve a switching off of $IL-2R^+$ cells (Smith *et al.*, 1984).

Our understanding of the reasons behind the known decline in immune function with increasing age is rather poor. It has been suggested that one of the main reasons for the deficiency in the immune response of the aged is due to a reduction in the number of cells which can be activated by lectins. It is also possible that the response to mitogenic signals of the lymphocytes from the elderly is inferior. Indeed, a comparison of the production of IL-2 and its receptor expression in human lymphocytes obtained from elderly and young donors has demonstrated that both IL-2 production and high affinity IL-2R generation are defective in the T cells from aged humans. Although the reasons for these defects are not understood at present, there can be little doubt that one of the main reasons for the decline of immune function in the elderly is, at least in part, due to the decrease in the capacity of lymphocytes to produce IL-2 and its receptor with the same efficiency and to the same extent as the corresponding cells obtained from young and healthy individuals (Froelich *et al.*, 1988).

The time sequence of events of the mitogenic stimulation of T lymphocytes has been probed further by testing the effects of cyclosporin A (Granelli-Piperno, Andrus & Steinman, 1986). When this compound is included in T cell cultures, a complete inhibition of IL-2 production occurs. However, even in the presence of cyclosporin A, concanavalin A stimulates all the early events of the signal transduction process, i.e. the hydrolysis of membrane phosphoinositides and the consequent generation of second messengers. The activity of protein kinase C is also unaffected by cyclosporin A and both these results suggest that cyclosporin A can inhibit only the later stages of signal transduction. As the combined effects of the phorbol ester and the calcium ionophore are known to bypass the early membrane events of mitogenic stimulation but still lead to T cell activation, the inhibition of IL-2 production by cyclosporin A and the total abrogation of the mitogenic stimulus under such conditions confirms the overriding importance of the late and not well-known events in the complex process of mitogenic stimulation (Mizushima *et al.*, 1987).

The effects of cyclosporin A have also been investigated in the *in vivo* popliteal lymph node model (Black *et al.*, 1988) where it has been confirmed that cyclosporin A inhibits the expression of the receptor for IL-2 *in vivo* as well as *in vitro*. The immunosuppressive agent blocks both spontaneous and IL-2-driven proliferation in concanavalin A stimulated cells, in contrast to the mitogenic stimulation achieved by allogeneic cells, suggesting that there are several possible pathways for T cell activation, some of which are independent of lymphokines (Black *et al.*, 1988).

One of the late stages in lymphocyte stimulation is characterized by increased glucose utilization of the cells. The tumour-promoting phorbol esters, such as 12-myristate 13-acetate (PMA), are not effective mitogenic stimulants when used alone. However, in combination with concanavalin A, PMA stimulates the synthesis of both DNA and protein to a significantly greater extent than that obtained with either of the mitogens used individually. PMA and concanavalin A together, but not individually, also increase the levels of the activity of the glycolytic enzymes in peripheral lymphocytes 48 h after the initial stimulation. This late timing of the increase in, for example, hexokinase activity suggests that gene expression for the glycolytic enzymes probably occurs in the S phase of the cell cycle. The expression of hexokinase is, thus, regulated by the cellular signalling system and is the result of the initial combined action of concanavalin A and PMA (Marjanovic, Wielburski & Nelson, 1988).

In addition to IL-2, the lectin-induced membrane reactions result in the production of other factors which also influence the mitogenic activation of T lymphocytes. One of the most studied of such factors is γ-interferon (Wheelock, 1965), and, because of its therapeutic application potential, the production of γ-interferon by activated lymphocytes has been studied extensively. The phorbol ester, TPA, when applied together with kidney bean lectins, PHA, has an interesting synergistic effect on the production of interferons by T cells, and the amounts of γ-interferon obtained from peripheral human leucocyte cultures can be reliably increased by the combined application of the two stimulants. In contrast, interferon production is somewhat variable and unpredictable on stimulation with PHA alone. It is also known that some T cell lymphoma and hybridoma lines can produce γ-interferon at maximal rates with the phorbol ester alone (Vilcek, Le & Yip, 1986). Furthermore, it is interesting that, although some lectins, such as PHA, induce the synthesis of interferons from some blood cells, others, such as concanavalin A, can inhibit the receptor-mediated internalization of recombinant interferons in those cells which express high affinity interferon α-receptors (Faltynek *et al.*, 1988). This, again, shows the great complexity of cell membrane events during lectin-induced mitogenesis.

T cell cultures stimulated with both PHA and phorbol ester produce cytokines such as lymphotoxin and monocyte cytotoxin (Stone-Wolff *et al.*, 1984). The presence of cytotoxins in the lymphocyte cultures may pose problems for the potential further use or therapeutic application of the γ-interferon produced. In lectin-stimulated lymphocyte cultures, several other toxic factors are also present and some of these can stimulate B lymphocytes (Kawano, Iwato & Kuramato, 1985) or granulocyte-macrophages (Cutler *et al.*, 1985). In addition, other less well-studied factors are also present, such as some natural inhibitors of lymphocyte proliferation and potential immunomodulators or immunosuppressors. A protein, $M_r = 61,000$, was shown to have such activity in normal chicken sera and it may be one of the prime examples of this class of factors (Davila *et al.*, 1987).

However, with the exception of IL-2, the role in lymphocyte culture of other stimulatory factors in the reaction mechanism of the mitosis is poorly understood.

Our understanding of the late events in the mitogenic stimulation of lymphocytes is rather poor, but it is known that protein kinase C participates in the phosphorylation of tyrosine residues in a number of both cytosolic and particulate cellular proteins after stimulation with lectins. Some, or all, of these may be similar to, or identical with the tyrosine residues of receptors phosphorylated in insulin-sensitive cells after insulin-binding to the membrane receptor. There is, however, no clear evidence so far to indicate which cellular proteins serve as substrates for these phosphorylation reactions.

Recent observations suggest two possible candidates for tyrosine kinase target proteins. One of these is a lipid kinase which phosphorylates phosphatidylinositol in the 3 rather than the 4 position. This is a target enzyme of oncogene-encoded and growth factor receptor tyrosine kinases (Courtneidge & Heber, 1987). The second candidate is a *cdc2* gene-encoded protein serine kinase which appears to be essential for starting up the cell cycle at G_1 phase and the entry of cells into mitosis (Draetta *et al.*, 1988; Michell, 1989). In addition, nuclear histones and other nuclear proteins are also likely substrates for the phosphorylation (Mire *et al.*, 1986*b*). In fact, it has been demonstrated by an *in vitro* gel assay system that extracts from concanavalin A-treated lymphocytes can readily phosphorylate histones (Grove & Mastro, 1987). The occurrence of similar reactions *in vivo* may lead to de-repression of DNA which may explain the enhanced level of RNA and protein synthesis in lymphocytes after stimulation with lectins. There is some experimental evidence to show that most cells respond to treatment with mitogenic lectins by an increased level of mRNA production (Messina, Hamlin & Larner, 1987*a*,*b*). This effect is similar to the observed increase in cellular protein synthesis in a number of eukaryotic cells after insulin-binding. Although lymphocytes have in the past been regarded as unreactive with insulin, it is now clear that they do respond to insulin stimulation if first activated with mitogenic lectins (Buffington *et al.*, 1986). It appears therefore, that increased mRNA production, and a consequent increase in the synthesis of some proteins with key functions in cellular metabolism and the cell cycle, is a general effect of most, if not all, proliferative signals.

Mitogenic lectins (but not incomplete mitogens, such as wheatgerm agglutinin) and insulin bind to their receptors on the nuclear envelope as well as on the plasmalemma and binding is followed by the activation of the receptor tyrosine kinase on both membranes (Purrello, Burnham & Goldfine, 1983*a*). Both insulin and concanavalin A also inhibit the incorporation of ^{32}P into nuclear membrane proteins (Purrello *et al.*, 1983*b*). It appears that a part of the initial stimulus may also be transmitted to the nucleus in addition to the effect of the mitogenic agent on the plasma membrane, leading to various membrane events and reactions and to the generation of second messengers. This is especially likely to occur with lectins which, at low concentrations, are taken up by cells preferentially (Sorimachi, 1984) and therefore, endocytosed lectins may directly affect the nucleus. Such specific binding of lectins to the nucleus of the sea urchin embryo has been shown experimentally in cells made permeable with glycerol treatment. Moreover, the amounts of fluorescein-labelled lectins can be quantitated in isolated chromatins. Changes in lectin-binding have been correlated with different stages of development of the embryo (Kinoshita, Yoshii & Tonegawa, 1988). There are also well-defined changes in the structure and protein composition of the nuclear matrix during concanavalin A-induced mitogenesis. Between 18 to 36 h after stimulation, the nuclei of T lymphocytes become totally restructured and undergo a sixfold increase in volume. There is a tenfold expansion of the interchromatin compartment due to continued protein synthesis. The proteins, RNA, and other components synthesized, accumulate in this compartment before transcription and DNA replication can take place. It is not clear whether such nuclear remodelling necessitates the transport of the primary signal, concanavalin A, to the nucleus (Bladon *et al.*, 1988) or whether the results are the effects of second messengers acting upon it.

In the last few years, interest in the selective transport of cell components through nuclear pores has been increasing. For example, it has been demonstrated that not all soluble cytoplasmic proteins may be transported through the nuclear pore into the nucleus proper, the main diameter of which is about 90 Å (Paine, Moore & Horovitz, 1975). Molecules of smaller diameter than 90 Å may enter the nucleus by passive diffusion, while larger molecules cannot, hence the composition of the nuclear pore is of great interest. Apparently, the main components consist of several glycoproteins which all bind to concanavalin A (Gerace, Ottaviano & Kondor-Koch, 1982), while other functional components, such as the pore glycoprotein of $M_r = 63,000–65,000$, isolated from rat liver nuclei and found to bind wheatgerm agglutinin, are also present. The binding of this lectin is specific and can be inhibited by N,N',N''-triacetylchitotriose (Davis & Blobel, 1986). Furthermore, by using an *in vitro* transport system of rat liver nuclei (and *Xenopus* egg extract), it has been shown that the nuclear transport of a fluorescently labelled nuclear protein, nucleoplasmin, is efficiently inhibited by wheat-germ agglutinin but not by other lectins. Moreover, after incubation with ferritin-labelled wheatgerm agglutinin, the cytoplasmic face of nuclear pores in rat liver nuclei contain large amounts of labelled lectin (Finlay *et al.*, 1987). Microinjection of wheatgerm agglutinin into living cells of *Xenopus* oocytes blocks the migration of proteins into the cell nucleus. Similarly, injection of fluorescently labelled nucleoplasmin into rat hepatoma cells, alone, or in combination with wheatgerm agglutinin, inhibits the uptake by the nucleus of karyophilic proteins regardless of size (Dabauvalle *et al.*, 1988).

The blockage by wheatgerm agglutinin of the active nuclear transport of nucleoplasmin in *Xenopus laevis* oocytes has been confirmed by several workers. In the same system, wheatgerm agglutinin also inhibits the nuclear transport of non-nuclear proteins conjugated with a synthetic peptide containing the nuclear localization signal sequence for simian virus 40 (SV40) large T antigen. The inhibition lasts for about an hour after the injection of the lectin and then gradually decreases. In contrast, the export of RNA from the nucleus into the cytoplasm is apparently not affected by wheatgerm agglutinin. Moreover, the lectin does not inhibit the passive diffusion of fluorescein isothiocyanate labelled dextran ($M_r =$ 17,900) into the nucleus. Other lectins, including *Wistaria floribunda* agglutinin, concanavalin A or lentil lectin are without effects on active nuclear transport. Thus, blockage by wheatgerm agglutinin of active (but not passive) protein import into the nucleus is a highly specific process (Yoneda *et al.*, 1987).

Lectins may also be involved in the control of transport from the nucleus into the cytoplasm. A mannose-specific lectin isolated from the coral, *Gerardia savaglia*, strongly inhibits the transport of mRNA through the nuclear pore, probably by binding to the glycoprotein components. Moreover, the coral lectin inhibits both mRNA efflux from isolated nuclei and the activity of nucleoside-triphosphatase (NTP-ase), which is the mediator of poly(A)-rich mRNA release. This is clearly a lectin effect as the inhibition of the mRNA efflux is abolished in the presence of mannose (Klajic *et al.*, 1987). It is of interest that concanavalin A, another mannose-specific lectin, binds to the nucleus only slightly in this system and does not appear to interfere with the nuclear-cytoplasmic mRNA transport. Concanavalin A binds mainly to the endoplasmic reticulum and components of the Golgi apparatus (Tartakoff & Vassali, 1983).

The existence of a direct interaction between lectins and the nucleus is undeniable, but the picture emerging from the complex and rather disjointed pieces of experimental evidence available is unclear. It appears to be well established that some lectins and, particularly that from *Gerardia*, affect the transport of cellular components between the nucleus and the cytoplasm. However, it is by no means certain that the nuclear events which precede mitosis are the result of direct effects of lectins on the nucleus and not due to the activities of second messengers generated by the original lectin-binding to the plasma membrane of the lymphocytes. It may be of special significance that, although the *Gerardia* lectin can cause a proliferative response from spleen cells, it can do it only in the presence of a lipopolysaccharide B cell mitogen (Reinherz *et al.*, 1979). No mitosis occurs with the *Gerardia* lectin alone, or even when it is combined with the T cell mitogen, concanavalin A.

In conclusion, during the transformation of resting lymphocytes into proliferating cells by stimulation with mitogenic lectins, some of the early changes in the sequence of events are now reasonably well known. Unfortunately, this applies much less to our understanding of the events of the later stages of the stimulation, which are known in main outline only. Nevertheless, it is possible to draw up a timetable, which may describe the main features of the changes leading up to mitosis (Figs 5.1 and 5.2).

Pre-treatment of lymphocytes with some enzymes may occasionally alter the extent of the mitotic response. Some lectins, such as the soyabean lectin, are not mitogenic for murine lymphocytes without first treating the cells with neuraminidase (Novogrodsky & Katchalski, 1973). However, enzymic removal of the galactose residue uncovered by the neuraminidase treatment, again renders the lymphocyte totally unresponsive to soyabean lectin (Novogrodsky *et al.*, 1975).

The state of polymerization and/or the valency of the lectin may also have a bearing on its effectiveness as a mitogen. For example, native soyabean lectin is a poor mitogen but after gentle polymerization, its mitogenic potency is increased, particularly for lymphocytes from which the terminal sialic acid residues have been removed. Similarly, the mitogenic effectiveness for rat lymphocytes of wheatgerm agglutinin and its succinylated derivatives is dependent on their state of oligomerization (Moullier *et al.*, 1986). Oligomerization of wheatgerm agglutinin may actually occur on lymphocyte membranes and it has been suggested that this is responsible for the tight co-operative binding of the lectin to its receptors. The molar concentration of monomeric concanavalin A required to achieve a maximum response with murine lymphocytes is

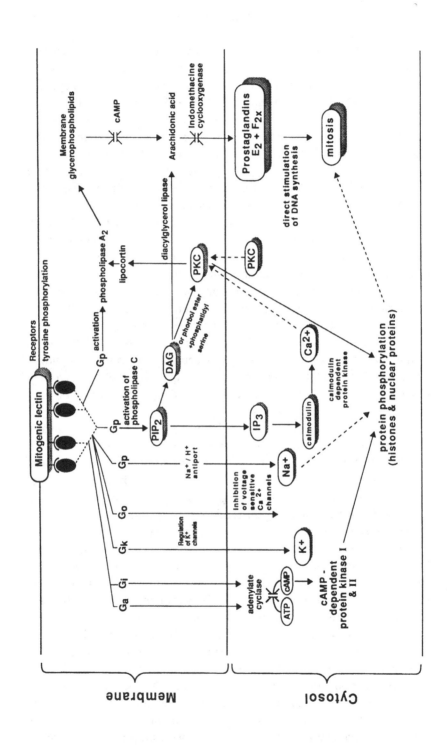

about 300 times higher than that of the native tetrameric lectin (Ishii *et al.*, 1984). Thus, the valency and/or the size of lectins appears to have a quantitative effect on the mitogenic response obtained with lymphocytes.

In some instances, reduction in the valency of lectins results in the expression of new biological activities not noticed with the parent, native tetrameric lectin. For example, succinylated or acetylated concanavalin A lyses sheep erythrocytes in the presence of guinea pig complement whilst the native lectin is inactive. The effect is specific for concanavalin A as, for example, succinylated wheatgerm agglutinin is inactive. Moreover, the haemolytic activity of succinylated concanavalin A is inhibited by α-D-methylglucopyranoside. The degree of lysis depends on the dose of the succinylated or acetylated derivative of concanavalin A and correlates with the fixation and activation of the first component of complement C_1 to the lectin-treated cells (Langone & Ejzemberg, 1981).

Immunosuppression

It has been recognized for some time that mitogenic stimulation of lymphocytes with lectins can produce suppressor cells which are capable of inhibiting all biological activities of both T and B cells. Apparently, suppressor cells may have an overall control of the functions of the entire immune system *in vivo* (Waldman & Broder, 1977) and suppressor cell dysfunction can induce the production of excessive amounts of autoantibodies. For example, such a dysfunction of immunosuppression in patients with Graves' disease may be the reason for the presence of increased concentrations of antithyroid autoantibodies (Ueki *et al.*, 1988).

Concanavalin A may have either stimulatory or inhibitory effects on the mitogenic response of mouse spleen cells to antigen stimulus (Dutton, 1972, 1975). Lymphocytes may generate suppressor cells in variable amounts depending on the concentration of the lectin, cell density, the presence or absence of serum in the culture, the age of the cell culture or the precise time of the introduction of concanavalin A into the medium in relation to that of the antigen. Moreover, cells previously exposed to concanavalin A may also become suppressive to fresh spleen cells.

Induction of the formation of suppressor cells by treatment with

Fig. 5.1. Receptor-couplings on mitogenic stimulation of resting (G_0) small peripheral lymphocytes. G proteins (see different subscripts for the various coupling processes); cAMP: cyclic AMP; IP_3: inositol 1,4,5, triphosphate; PIP_2: phosphatidyl inositol 4,5 diphosphate; DAG: diacylglycerol; PKC: phospholipid-dependent protein kinase C.

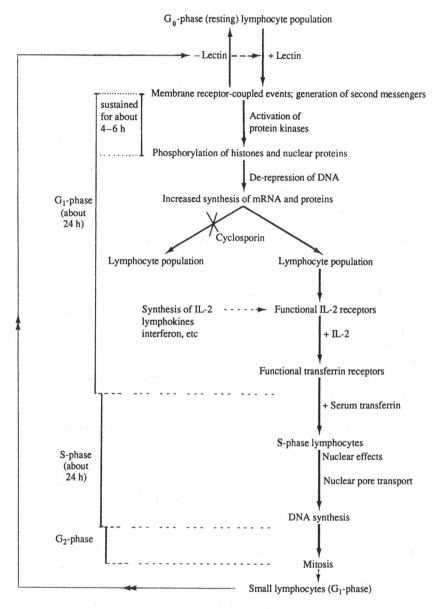

Fig. 5.2. Approximate time-table during mitosis of resting (G_0) lymphocytes. IL-2: interleukin 2.

concanavalin A occurs in human spleens and treatment with PHA has also produced cells which can suppress lymphocyte transformation in mixed human lymphocyte cultures (Sampson, Grotelueschen & Kauffman, 1975). Moreover, supernatants obtained from mouse spleen cells stimulated with concanavalin A can mediate the generation of suppressive effects in fresh cell cultures (Rich & Pierce, 1974). In fact, after activation with lectins, several soluble glycoprotein factors which mediate the effects of suppressor function have been isolated from lymphocyte cultures. Thus, the transformation inhibitory factor for T lymphocytes, obtained after mitogenic induction with specific antigens or lectins, is a glycoprotein of $M_r = 30,000$ to $45,000$ (Greene, Fleischer & Waldman, 1981) whose effects are abolished in the presence of N-acetyl-D-glucosamine. A similar factor of $M_r = 60,000$ to $80,000$, inhibitory for B cells, has been isolated from supernatants of B cell cultures stimulated with pokeweed mitogen. However, the activity of this B cell suppressor factor can be abolished by a different carbohydrate, L-rhamnose (Fleischer *et al.*, 1981). Other similar soluble B cell suppressor factors have also been obtained but it is not clear what relationship, if any, exists between them. For example, the smaller soluble factor isolated by Warrington *et al.* (1983) with $M_r = 20,000$ to $25,000$ appears to be distinctly different from the one isolated previously (Fleischer *et al.*, 1981), although they can both be inhibited by the same sugar, L-rhamnose. The observation that the inhibition of the activities of these factors is carbohydrate specific does suggest that they may all be lectins.

Macrophages treated with soluble suppressor factors can produce another suppressor factor after about 1 to 3 days. This is a protein of 50,000 to 55,000 molecular weight which can suppress most immune responses *in vitro*, including suppression of plaque forming responses, the mitogenesis of both B and T cells and the synthesis of antibodies (Aune & Pierce, 1981*a*, *b*).

In practice, however, considerable difficulties are encountered with the *in vitro* culturing of suppressor T cells since there are some indications that the molecular signals needed for their stimulation may be different from those required for the culturing of other T lymphocytes. It has been found that cultures of peripheral blood mononuclear cells stimulated with pokeweed mitogen contain an excess of T8[+] cells which suppress the mitogenic response of fresh T4[+] cells to pokeweed mitogen. This response is lost if T4[+] are removed from the original blood cell culture stimulated by pokeweed mitogen. Accordingly, it appears that the *in vitro* stimulation of helper cells with pokeweed mitogen produces a suppressor cell growth factor; a small protein of $M_r = 8000$, which has been purified by high pressure liquid chromatography. It is not identical with IL-1, IL-2 or γ-interferon and its main function is apparently to induce IL-2 receptor expression on T8$^+$ suppressor cell precursors (Fox *et al.*, 1986).

Recent studies with highly purified T cell subsets have further emphasized the complexity of the activation reactions of resting pure T4$^+$ and T8$^+$ cells and have also pointed up the differences in the stimulation of helper and suppressor subsets of T lymphocytes. Thus, while T4$^+$ cells are capable of mounting a proliferative response to allogeneic non-T cells on their own, T8† cells require 'help' from either autologous helper cells or from exogenously supplied IL-2. Accordingly, suppressor cells are unable to produce their own IL-2 under some experimental conditions and, therefore, they cannot start DNA synthesis and will not proliferate (Halvorsen *et al.*, 1987, 1988). This contrasts with the findings of other studies in which T8$^+$ cells have been shown to produce IL-2 and to proliferate in the presence of accessory cells (Meuer *et al.*, 1982; Lea *et al.*, 1986). In highly purified subsets obtained by immunomagnetic separation, when no antigenic stimulus is applied, both T4$^+$ and T8$^+$ cells give comparable responses to stimulation by PHA in the presence of accessory cells. In their absence, however, PHA induces low levels of IL-2R expression and both subsets also show definite, although low level, responsiveness to IL-2. If accessory cells or 12-*o*-tetradecanoyl-phorbol-13 acetate (TPA) are also present in the culture, both subsets of T cells can produce IL-2 and synthesize DNA. Short preincubation with PHA 'primes' the T cells and they respond well to subsequent application of either IL-2 or TPA, while the opposite order of adding the reagents does not stimulate IL-2 production. Thus, PHA alone activates IL-2R expression in T cells. However, IL-2 production can occur only through the direct interaction of PHA and accessory cells. Accordingly, while T8$^+$ cells stimulated to proliferate with antigen require the help of T4$^+$ cells, in PHA stimulated lymphocytes, the requirements for activation and proliferation appear to be identical for both helper and suppressor cells (Leivestad *et al.*, 1988). The differences between the behaviour of T cell subsets to stimulation by antigen or PHA and the requirements for the presence of other factors can be visualized from the schematic diagram (Fig. 5.3).

Another paradox, which is not completely understood at present concerns the behaviour and role of monocytes in the mitogenic response since they may either enhance or suppress the mitogenic response of human T lymphocytes to different lectins under identical culture conditions. Thus, *Datura stramonium* lectin is less mitogenic when mono-

T^{4+} + Antigen ----> IL-2 + T^{4+}-IL-2R ----> DNA synthesis;
proliferation

T^{8+} + Antigen ----> IL-2 + T^{8+}-IL-2R + AC (or IL-2) ----> DNA synthesis
proliferation

T^{4+} ----> T^{4+}-IL-2R   [no IL-2 production]+ TPA ----> DNA synthesis
  + PHA   [no DNA synthesis]or AC
T^{8+} ----> T^{8+}-Il-2R   [no proliferation   ]+ TPA ----> DNA synthesis

T cells + TPA ----> no activation + PHA ----> no DNA synthesis

Fig. 5.3. Schematic representation of the stimulation of T cell subsets by antigen or PHA. IL-2: interleukin 2; IL-2R: interleukin 2-receptor; T^{4+}: 'Helper T cells'; T^{8+}: 'Suppressor T cells'; TPA: tetradecanoyl-phorbol acetate; AC: exogenous accessory cells.

cytes are included in the cell culture whilst the mitogenic response of the same lymphocytes to wheatgerm agglutinin is appreciably increased by the presence of these accessory cells (Kilpatrick, 1988). Although it is not likely to be the result of cytotoxicity, it is possible that the inhibition with the *Datura* lectin is cell mediated. These findings again draw attention to the difficulties of comparing results obtained in different laboratories, particularly if the experiments of mitogenic stimulation have been carried out in the presence of a range of accessory cells.

Suppression of humoral and cellular immunity is usually ascribed to T suppressor cells (Basten *et al.*, 1974), macrophages or monocytes (Becker *et al.*, 1981; Rice, Laughter & Twomey, 1979), but rarely to B cells (Koenig & Hoffman, 1979). However, in highly purified B cell populations obtained from human peripheral blood, suppressor B cells are induced by PHA. The generated B suppressor cells directly inhibit allogeneic mixed lymphocyte reactions and PHA-induced DNA synthesis. These effects are not mediated through the induction of T suppressor cells (Farkas, Manor & Klajman, 1986). It is envisaged that suppressor B cells may have important, although as yet unknown, immunoregulatory functions *in vivo*.

The immunosuppressive effects found *in vitro* with lymphocyte cultures stimulated with lectins are usually implied to occur *in vivo* but the demonstration of such immunosuppressive effects in experimental animals is much more difficult to achieve than in cell cultures. For example, the plaque-forming cell response can be suppressed in mice given concanavalin A for 4 to 24 h. However, spleen cells obtained from these mice have no suppressive activity when transferred to normal recipients who respond only if they are also given concanavalin A within 48 h of cell transfer and immunization. Moreover, spleen cells obtained from mice pre-treated with concanavalin A for 4 to 24 h, can suppress the primary antibody response to sheep erythrocytes *in vitro*. The results have been interpreted to show that a suppressor cell population is activated in the animals, following the *in vivo* treatment of mice with concanavalin A, but that this activation is lost when the lectin is removed (Smith *et al.*, 1984).

Lectin-induced immunosuppression has important applications in medical–clinical practice. The measurement of suppressor T cell function can give diagnostically useful indications about the immune competence of patients suffering from diseases caused by deficiencies in the immune system (Armitstead & Ewan, 1984). Even more important is the degree of tolerance induction by lectins observed in allograft recipients (Calne, Wheeler & Hurn, 1965; Markley *et al.*, 1969; Hilgert *et al.*, 1980; Harel, Banai & Nelken, 1981). Unfortunately, as it is commonly found, such immunosuppression does not last long *in vivo* without the continuous administration of the lectin (Smith *et al.*, 1984). Despite this constraint, the survival of skin allografts can be extended significantly over that of controls by pre-treatment of the recipients with PHA for about a week before transplantation, followed up by giving small doses of PHA on alternate days after transplantation. The survival time of the allograft can be even further extended if both the donor and the recipient rats have been pre-treated with PHA (Ludwin & Singal, 1986). Similar findings have been made with lentil lectin which is claimed to be the most effective suppressant of graft versus host reaction (Hilgert *et al.*, 1980; Harel *et al.*, 1981). Moreover, while the administration of non-toxic amounts of lentil lectin to both donors and recipients of allogeneic spleen cells suppresses the development of regional and systemic graft versus host reaction, the lectin treatment does not appear to cause damage to the stem cells (Sula *et al.*, 1986). Administration of PHA also inhibits the development of thyroglobulin-induced autoimmune thyroiditis provided it is injected intravenously before the exposure to thyroglobulin (Esquivel, Mena & Folch, 1982). It is possible that, when immunogenic food proteins are introduced orally with PHA, suppression of antibody production is also due to the general immunosuppressive effects of lectins (Pusztai *et al.*, 1981*b*; Grant *et al.*, 1985).

Lectin-dependent blood cell cytotoxicity and phagocytosis

When lectins or other mitogenic agents are present in the cultures of blood cells, the killing of target cells by effector cytotoxic cells is independent of the presence of antigens. Thus, lectins may abrogate the interaction between effector and target cells based on the recognition of the major histocompatibility complex antigens. In lectin-dependent cytotoxicity reactions, the first step is a cross-link formation by the lectin between effector and target cells (Asherson & Ferluga, 1973). This is followed by the activation of the effector cells. Due to the cross-linking and the close proximity of the two cells, the target cells may be killed by the cytolytic activity of the activated effector cells (Green, 1982). Alternatively, the reaction of the lectin with membrane glycoconjugates may convert potential cryptic structures on the target cells into membrane components which are more reactive towards the effector cells and thus facilitate their ultimate lysis (Bonavida & Katz, 1985). The complexity of the reaction is shown by the observation that concanavalin A may either induce or inhibit cell-mediated lympholysis in murine lymphocytes (Beretta *et al.*, 1982; Sitkovsky, Pasternack & Eisen, 1982). It appears that the reason for this paradoxical behaviour of the lectins is due to their direct triggering effects on the lytic potential of activated T cells. Thus, concanavalin A-coated target cells are efficiently lysed by cytotoxic cells but the addition of soluble concanavalin A to the system leads to a dose-dependent inhibition of the lysis. As the same inhibition can also be achieved by the addition to the culture of a monoclonal antibody specific for T cell lines instead of concanavalin A, this inhibition appears to be exerted at the effector cell level. On the other hand, concanavalin A has no clear inhibitory effect on the cytotoxicity exerted by freshly isolated peripheral mononuclear blood cells against the same target cells. The results suggest that the lectin may directly trigger the lytic activity of the effector cells. Accordingly, lysis of target cells occurs when the coating of these cells with concanavalin A triggers the cytolytic activity of effector cells. Although soluble concanavalin A also triggers the lytic machinery of effector cells, the resulting release of soluble cytolytic factors leads to the production of exhausted effector cells which are no longer capable of target cell lysis (Miltenburg *et al.*, 1987).

Under somewhat similar conditions, some lectins, like wheatgerm agglutinin, may induce the killing of tumour cells by macrophages (Kurisu, Yamazaki & Mizuno, 1980). This reaction bears some resemblance to macrophage-mediated tumour cell lysis by anti-tumour antibodies. Moreover, the mechanism of the lectin-mediation in the reaction between macrophages and tumour cells is probably similar to that occurring with cytotoxic effector T cells in lectin-dependent cytotoxicity reactions. The first step in the reaction sequence, similar to that occurring in lectin-dependent cytotoxicity reactions, is the cross-linking of carbohydrate receptors on the membranes of both macrophages and the tumour cells (Kurisu *et al.*, 1980). Both types of cytotoxic reactions may occur *in vivo*. For example, the protection of mice from tumour development on inoculation with Ehrlich ascites cells by their pre-treatment with *Griffonia simplicifolia* lectin, may be the result of such lectin-mediated cytotoxic reaction (Eckhardt, Malone & Goldstein, 1982).

Lectins can also facilitate the phagocytosis of target cells. Thus, administration of additional lectin to rosettes formed between concanavalin A-coated murine macrophages and erythrocytes, activates the macrophages which then phagocytose the erythrocytes (Sharon, 1984). This phagocytosis may be the major non-opsonin-dependent type occurring *in vivo*, particularly as it applies to the phagocytosis of bacteria (Sharon, 1984; Ofek & Sharon, 1988). Most studies into the nature and reaction mechanism of bacterial lectinophagocytosis have been carried out with extraneous plant lectins. However, the reaction which occurs *in vivo* is caused either by surface lectins from bacteria binding to carbohydrate structures on macrophage membranes or by endogenous membrane lectins from macrophages recognizing and binding to bacterial surface carbohydrates. Thus, the existence of such a mechanism *in vivo* needs further verification.

Lectins can have dual effects on macrophages. Some may potentiate bacterial uptake by macrophages to suppress their bactericidal activity. For example, pre-treatment of murine bone marrow-derived macrophages with wheatgerm agglutinin or concanavalin A markedly enhanced the phagocytosis of *Staphylococcus aureus* H and *Escherichia coli* 08. In contrast, the same treatment suppressed or completely abolished their bactericidal activity against the same two microorganisms (Gallily, Stain & Zaady, 1986).

Lectin-induced mast cell reactions

Basophil and/or mast cells release histamine and/or other pharmacologically active amines on exposure to a number of secretagogues and this is the basis of allergic, immediate or Type-1 hypersensitivity reactions.

Under certain conditions, and in predisposed individuals, the first step of the classical allergenic response to antigens is the synthesis of a specific IgE type antibody, which binds to the surface of mast cells. On re-exposure, the sensitizing antigen binds to this surface IgE antibody and cross-links the individual mast cells. Free Ca^{2+} ions are subsequently liberated in the cytosol in a reaction which is not fully understood but which bears a strong resemblance to that occurring in lymphocytes during mitogenesis. Observations made with calcium antagonists of a cloned murine mast cell line, MC9, after stimulation with concanavalin A have suggested a role for a voltage-independent but Ca^{2+}/calmodulin-dependent plasma membrane Ca^{2+} channel in the accumulation of calcium ions in the cytosol. In this process the MC9 cells release Ca^{2+} from intracellular stores via hydrolysis of membrane phosphoinositides and the released calcium forms a complex with calmodulin to activate the Ca^{2+} channel. Due to the increased entry of calcium, the concentration of free cytosolic Ca^{2+} is elevated, which is followed by histamine release. It is not known if the activation of this channel requires phosphorylation or not (Gulbenkian *et al.*, 1987). Thus, although the fine details of the reaction mechanism of mast cell activation after stimulation with lectins has not yet been settled, the final outcome of the stimulation is not in doubt. This energy-dependent, exocytotic reaction of histamine secretion, which is the main biological function of mast cells, occurs without any damage to the cells.

Lectins are potent stimulants of the degranulation of mast cells (Siraganian & Siraganian, 1974; Hook, Dougherty & Oppenheim, 1974; Bach & Brashler, 1975; Pierce, 1981; Lima *et al.*, 1983) and lectins, such as concanavalin A, may also be capable of stimulation of the synthesis of histamine in cells totally unrelated to mast cells (Nolte, Skov & Loft, 1987). For example, various sarcoma cells respond to stimulation with concanavalin A by histamine secretion but the role of other agents, such as Ca^{2+}, phosphatidylserine, etc, which have been suggested to facilitate exocytosis, is not clear (Truneh & Pierce, 1981; West, 1983; Horigome *et al.*, 1986).

Lectins induce degranulation in mast cells obtained from either sensitized or non-sensitized animals. Lectin-induced exocytosis is therefore not dependent on immune reactions. Moreover, when applied in concentrations which differ from those used in lectin-induced degranulation, lectins can inhibit exocytosis induced by other secretagogues (Bach & Brashler, 1975). Thus, lectins can bind to both IgE receptors of mast cells and to the carbohydrate side-chains on the immunoglobulin (IgE) molecule (Helm & Froese, 1981).

It is of special interest that lectins incorporated in diets are known to affect gut mast cells which immediately and directly respond to dietary PHA (Greer & Pusztai, 1985). Moreover, the anaphylactic response to later oral challenges with PHA is elevated after longer exposures to dietary PHA. Accordingly, it appears that the amplified response of mast cells to PHA in longer term experiments may be the result of the generation of gut mast cells which carry lectin-specific, IgE-type antibodies. The re-introduction of PHA into rat diets after feeding the animals on lectin-free diets for a period of 5 to 6 days, results in an elevated anaphylactic reaction, suggesting that there is memory effect in the local (gut) IgE-based immune system (Greer & Pusztai, 1985).

In mice, in addition to local (gut) hypersensitivity reactions caused by exposure to dietary PHA, systemic IgE-type antibodies are also produced and these animals develop a systemic immediate (Type-1) hypersensitivity reaction (Pusztai *et al.*, 1983*b*). Similar observations have been made with orally applied jack fruit, *Artocarpus integrifolia*, lectin in DBA/2 mice (Restum-Miquel & Prouvost-Danon, 1985). Depending on the amounts applied, a number of lectins, such as PHA or jacalin, may have different effects on the IgE antibody production to simultaneously injected but unrelated antigens. Simultaneously injected lectins may also produce similar effects on the development of other types of antibodies. There is some evidence to suggest that jacalin stimulates the production of helper IgE-producing lymphocyte function, while at higher doses it may be strongly immunosuppressive on IgE production (Lima & Pusztai, unpublished observations), raising the possibility that jacalin may be usefully employed in pharmacological practice. Hopefully, by interfering with the synthesis of IgE antibodies specific for potentially allergenic antigens, jacalin may possibly be used to reduce troublesome allergic responses to other antigens.

6

General effects on animal cells

Insulin-mimicking effects on adipocytes and other cells

The finding that lectins, such as concanavalin A, wheatgerm agglutinin, PHA and others, can mimic the effects of insulin on adipocytes is one of the most dramatic examples of interactions between lectins and cell membranes. Thus, these lectins stimulate the synthesis of triglycerides and the transport and oxidation of glucose in fat cells *in vitro* (Cuatrecasas & Tell, 1973; Pusztai & Watt, 1974; Livingstone & Purvis, 1980). Similar effects have also been observed *in vivo* (Katzen *et al.*, 1981).

It is clear that, depending on their carbohydrate specificities, various lectins display significant differences in their insulin-mimicking activities. These differences become clearer if the insulin effect is not tested by the commonly used glucose oxidation assay, but rather by assessing their effects on lipogenesis or from their inhibition of lipolytic effects of hormones, such as epinephrine or other lipolytic hormones, on rat or hamster adipocytes (Ng, Li & Yeung, 1989). Rat cells lack functional α-adrenergic receptors, while hamster adipocytes possess both α- and β-adrenergic receptors (Carpene, Berlan & Lafontan, 1983) and these variations help to differentiate between the insulin effects of the different lectins. Thus, the D-mannose and N-acetyl-D-glucosamine-specific lectins generally have both potent antilipolytic and lipogenic activities while the fucose-binding lectins possess antilipolytic activity only. With some exceptions, most of the D-galactose-specific lectins show slight insulin-like effects. Furthermore, the N-acetyl-D-galactosamine-specific *Dolichos biflorus* and *Maclura pomifera* lectins exert antilipolytic activity in hamster but not in rat adipocytes. Of those studied, the lectin from *Bauhinia purpurea*, specific for D-gal(1–3)-N-acetyl-D-galactosamine, is exceptional in that it shows potent lipogenic activity, although it has no antilipolytic activity (Ng *et al.*, 1989). Significantly, several lectins, such as the agglutinins isolated from soyabeans, peanuts, *Bandeirea simplicifolia* or *Sophora japonica*, neither bind to insulin receptors nor do they exhibit insulin-mimicking activities (Hedo, Harrison & Roth, 1981).

The reaction mechanism of the effects of lectins on adipocytes has not been studied in the same comparable detail to similar studies on the growth hormone-induced changes in lipid metabolism. It is, therefore, not known if the lipolytic effect of lectins can be blocked by inhibitors of protein and RNA synthesis or inhibitors of Ca^{2+} uptake in a way similar to that found in adipocytes stimulated with growth hormone. The possible blockage of the antilipolytic effects of lectins on adipocytes by α-difluoromethylornithine (DFMO), an irreversible inhibitor of the *de novo* biosynthesis of putrescine by ornithine decarboxylase has not been tested (Campbell & Scanes, 1988) nor is it clear whether the inhibition of the antilipolytic effects of lectins can be reversed by the administration of spermidine. Thus, the involvement of second messengers in the reversal of lipolytic hormone-induced lipolysis by lectins has not yet been shown conclusively, although the main determinants of biological activity in the fat cell system are the carbohydrate residues of the adipocyte membrane receptors for both hormones and/or lectins. However, the precise location, configuration and other structural features of the receptors binding the hormones or the lectins may differ sufficiently to explain the differences in the reactivities of the different lectins tested (Ng *et al.*, 1989).

In addition to binding to the same or similar receptors on the adipocytes (Cuatrecasas & Tell, 1973; Cuatrecasas, 1973), some lectins also mimick the effects of insulin on the activation or induction of various enzymes. Thus, lectins are just as effective as insulin in inducing tyrosine aminotransferase activity (Smith & Liu, 1981) and they also mimic the effects of insulin on the nuclear envelope phosphorylation (Purrello *et al.*, 1983*a*).

Most importantly, both concanavalin A and insulin stimulate the tyrosine kinase activity of the purified insulin receptor (Roth *et al.*, 1983). As previously discussed, both insulin and mitogenic lectins activate cells by binding to the same or very similar cell surface receptors and by inducing the generation of potent second messengers for transmitting the original signal to the inside of the cell. As a result of signal transduction, the cells begin to produce specific mRNAs in response to the original signal (Messina *et al.*, 1987*a*). However, there are also differences between the effects of lectins and insulin. For example, in rat H4 hepatoma cells, both concanavalin A and wheatgerm agglutinin appear to produce more mRNA than insulin. Also, the stimulation by lectins is of longer duration than that with insulin. Although the reasons for this quantitative difference between the stimulation achieved with insulin and lectins are not fully understood, this may be the result of the different behaviour of the two protein signals after the initial binding. In most cells insulin is endocytosed together with its receptor very soon after binding. Most of this complex is directed to the lysosomes, where insulin is degraded, while the receptor is re-cycled to the membrane of the cell. For sustained stimulation, therefore, further insulin molecules need to bind to membrane receptors. In comparison, the binding of some lectins to the receptor is longer lasting and to stop the stimulation, the bound lectin has to be released from the receptor by washing the cells with solutions of the appropriate specific carbohydrate hapten. Thus, a number of lectins may produce more persistent insulin-like effects in some cells than those obtained by stimulation with insulin (Schechter, 1983). Although the lectins are also ultimately endocytosed by cells, most likely by a receptor-mediated internalization, insulin receptors do not appear to take part in the intracellular processing of the lectins.

Despite the insulin-like activities of lectins *in vitro*, most observations indicate that *in vivo* both dietary PHA (Pusztai, 1986*a,b*; Pusztai, Grant & de Oliveira, 1986) and, to a lesser extent, soyabean agglutinin (Grant *et al.*, 1987*a*) reduce blood insulin concentration soon after reaching the small intestine. The effects on systemic metabolism of feeding rats with diets containing PHA, i.e. the elevated levels of fat, carbohydrate and protein catabolism, are most readily explained by such a drop in the concentration of circulating insulin. However, it is not clear how the dietary lectin contained in the lumen of the small intestine can modulate the concentration of a systemic hormone such as insulin.

Effects of lectins on cells of the digestive tract

Most plant-based foods are known to contain lectins because they are one of the main functional components of seeds and vegetative tissues. Arguably, one of the most important interactions between plants and animals *in vivo* occurs in the alimentary canal of animals ingesting foods of plant origin. As plants are the ultimate source of nutrition for the whole living world, the surface cells of the entire digestive tract must have been regularly exposed to lectins of all kinds during evolution. Moreover, lectin–epithelial cell interactions may have had an important role in the development of the digestive/absorptive function of the small intestine. Obviously, both humans and animals have learnt, mainly by trial and error, which foods are nutritious and which to avoid. Some plants are very highly toxic, such as the seeds of the *Ricinus* plant and these are largely avoided by both humans and animals, although accidental poisoning with such plants is known to occur (Pusztai, 1986*a*), in practice, this is rare. Despite this, studies of the effects on the alimentary tract of such highly toxic lectins are very important but the main thrust of the experimental investigations into the role(s) and function(s) of lectins in nutrition is directed towards the effects of those found in the traditional or more recently accepted common food plants.

Lectins in foods

The presence of lectins in foods has not been systematically investigated although there are several *ad hoc* surveys. However, most edible plants are known to contain some lectins and some have substantial amounts. For example, Nachbar and Oppenheim (1980) have found that about 30% of fresh and, more surprisingly, processed foods tested by them contain lectins. By measuring the haemagglutination activities of extracts prepared from green salads, fruits, spices, seeds, dry cereals and nuts, some even after roasting, they have demonstrated that most of such foods contain significant amounts of lectins. Their survey of the literature has also revealed another 53 edible plants with appreciable lectin activity. Some of these not only react with red cells used in the haemagglutination assays, but they also interact with serum or salivary components and bacteria from the oral cavity, such as *Streptococcus mutans* (Gibbons & Dankers, 1981).

In a more systematic survey (Grant *et al.*, 1983), the haemagglutination activity of legume seeds generally available in the UK against a number of human and animal red cells, a clear correlation between nutritional value of the legume seeds and their haemagglutination activity has been found. The more toxic the seeds are, the more extensively their lectins react with red cells. There are now several general methods available for the screening of the lectin content of common foods. Andersen and Ebbesen (1986) have recently described two very useful methods; one of these is based on line-dive immunoelectrophoresis, while the second is a haemadsorption test. Both methods have been tested extensively with all types of foods. A latex–haemagglutination technique described by Pongor and Riedl (1983) may also be adapted to screening tests. ELISA methods have also been developed for the determination of lectins with potential applications for food testing, for example, in the measurement of soyabean agglutinin in animal feeds by Prince *et al.* (1987). Similar immunochemical or functional assays have been presented by others.

All these studies have confirmed the presence of nutritionally significant amounts of lectins in all kinds of foods. Although heat treatment reduces the lectin content of some foods (Grant *et al.*, 1982; Pusztai, 1985), this is expensive and kept to a minimum in animal nutrition. However, some lectins are slightly affected by heating whilst others eaten by man are not heated at all. Thus, the gut of both humans and animals is continuously exposed to exogenous dietary lectins (Nachbar & Oppenheim, 1980; Pusztai, 1986*a,b*; Pusztai *et al.*, 1986; Liener, 1986).

Obviously, the biological responses to different dietary lectins are first and foremost dependent on the degree of their resistance to gut proteolysis. In addition to survival during gut passage, the carbohydrate specificity of the different lectins will determine which gut cells and tissues they may react with. Although in the last 20 years the effects of plant lectins on the digestive tract have been studied extensively, most of these studies have been confined to a few selected lectins. These have included concanavalin A, PHA, wheatgerm agglutinin or, more recently, the lectin obtained from the economically and nutritionally promising tropical winged bean, *Psophocarpus tetragonolobus*, plant. The effects of these lectins are generally detrimental to the growth and health of animals and some of them are overtly toxic. In contrast, there is much less known about the effects on the gut of potentially beneficial lectins, such as the tomato lectin (Kilpatrick *et al.*, 1985). Despite the selective nature of these studies, some firm and generally applicable conclusions relating to the effects of lectins on the functioning and structure of the gut have emerged. It has been confirmed that most lectins show appreciable resistance to gut proteolytic enzymes (Table 6.1). The proportion of reactive lectin remaining in the gut lumen is comparatively high in most instances although the degree of resistance may vary from lectin to lectin (Pusztai, 1986*a*; Pusztai *et al.*, 1986; Banwell *et al.*, 1983; Hara *et al.*, 1984; Kilpatrick *et al.*, 1985; Grant *et al.*, 1985). As expected, the surviving lectins are fully reactive and, depending on their carbohydrate specificity, they combine with appropriate components of the digesta, secreted, dietary or bacterial glycoconjugates and/or most importantly, with the membranes of gut epithelial cells. The dietary lectins interfere with the normal functioning of the alimentary tract through interactions with the functional components of the digestive system. In this process lectins may introduce changes in some, or all, of the digestive, absorptive, protective or secretory functions of the whole digestive system and affect cellular proliferation and turnover. Accordingly, it is recognized that lectins constitute one of the main physiologically active components of foods of plant origin and as some are antinutrients, the consumption of

Table 6.1. *Survival and binding to the rat small intestinal epithelium of intragastrically administered pure lectins.*

Lectins	Specificity	Binding	% Immunoreactive lectin recovered
PHA (*Phaseolus vulgaris*)	Complex	+++	>90
Con A (*Canavalia ensiformis*)	Man/Glc	++	>90
GNA (*Galanthus nivalis*; snowdrop)	Man	–	>90
SNA-I (*Sambucus nigra*; elderberry)	α-2,6-neuraminyl-Gal	+	50–60
SNA-II (*Sambucus nigra*; elderberry)	GalNAc	+++	>60
SBL (*Glycine max*; soyabean)	GalNAc/Gal	++	40–50
LEL (*Lycopersicon esculentum*; tomato)	GalNAc	+	40–50
WGA (*Triticum vulgare*; wheatgerm)	GlcNAc	+	50–60
PSL (*Pisum sativum*; pea)	Man/Glc	±	30–40
VFL (*Vicia faba*; broad bean)	Man/Glc	±	20–30
DGL (*Dioclea grandiflora*)	Man/Glc	±	18–20

Note: The strength of binding to the small intestinal mucosa is estimated from the immunoreactive PAP staining (see Fig. 6.21, p. 168) and expressed on an arbitrary scale between +++ = very extensive binding and – = no binding at all.

foods containing them can have serious consequences for growth and health.

Most of our knowledge concerning the effects of dietary lectins has been obtained from studies carried out with relatively few lectins, the most studied of these is the family of isolectins obtained from the seeds of kidney beans, *Phaseolus vulgaris*, PHA. As this lectin occurs in relatively high concentrations in bean seeds, diets containing significant amounts cause striking gastroenterological and other physiological effects in animals. It is therefore not surprising that our understanding of the antinutritional effects of PHA is more highly advanced than that of any similar interaction between the gastrointestinal tract and other legume lectins. Most of the early studies have been reviewed extensively by Jaffé (1980) whose pioneering work contributed largely to our initial understanding of the antinutritive properties of PHA. Several reviews have been published in the last ten years which update the information derived from pre-1980 studies with more recent developments (Pusztai *et al.*, 1982; Pusztai, 1986*a,b*; Pusztai *et al.*, 1986; Pusztai, 1989*a,b,c*).

Despite a general interest in lectins, most of our understanding of their toxic, gastroenterological effects in diets is the result of studies carried out with PHA. However, the results of more fragmentary studies with other dietary lectins largely support the findings obtained with PHA and the basic features of the interactions between different lectins and the digestive system are sufficiently similar as to allow us to draw some generalized conclusions.

The nutritional toxicity of PHA

It has been known since the turn of the century that inclusion of raw kidney bean seeds in the diets of monogastric animals leads to rapid weight loss and other similar manifestations of toxic effects (Fig. 6.1). In fact, at high enough concentrations, feeding animals on kidney bean-containing diets may result in death within a few days whilst properly heat-treated seeds are not toxic. Denatured kidney bean seed proteins support growth within the constraints of their deficiencies of essential amino acid content (Johns & Finks, 1920*a,b*).

A connection between the haemagglutinin component discovered in kidney bean seeds at the beginning of the century (Landsteiner & Raubitschek, 1908) and the toxic effects observed in animals fed on kidney bean-containing diets was suggested by Luning & Bartels in 1926 (for other early work, see Jaffé, 1980). More recently the seed lectin has been unequivocally established as the toxic factor (Jaffé, 1980; Pusztai & Palmer, 1977; Pusztai *et al.*, 1981*b*) and a clear dose–effect relationship

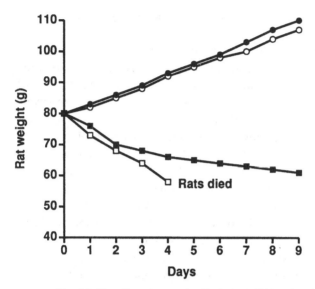

Fig. 6.1. The effects on growth of inclusion of kidney bean proteins in the diets of rats (weight of rats fed on different diets is plotted for 9 days). Open squares: diet containing 10% (w/w) kidney bean proteins; open circles: diet containing 10% (w/w) heated kidney bean proteins; filled circles: diet containing 10% (w/w) lactalbumin; filled squares: protein-free diet.

between dietary PHA concentration and the extent of the toxicity has also been shown (Pusztai *et al.*, 1981*b*; de Oliveira, Pusztai & Grant, 1988; Lafont *et al.*, 1988). PHA is toxic both for growing and mature rats (Grant *et al.*, 1985). All animal species tested so far, including ruminants (Williams *et al.*, 1984), birds (Andrews & Jayne-Williams, 1974) and humans (Griebel, 1950; Rainer, 1962; Haidvogl, Fritsch & Grubbauer, 1979; Bender & Reaidi, 1982) are known to be affected.

Although it is not entirely clear yet how PHA exerts its toxic effects, the main feature is that the rats which consume diets containing kidney bean lectin cannot properly utilize the protein content of the diet and, at the same time, have a highly elevated rate of tissue catabolism (Pusztai *et al.*, 1981*b*). Due to the poor utilization of dietary proteins combined with elevated breakdown of tissue proteins, the overall loss of protein from the body may exceed that taken in by the diet at or above sufficiently high dietary concentrations of PHA. Consequently, such animals are in a negative nitrogen balance (Pusztai *et al.*, 1981*b*). Thus, the primary effects of the dietary lectin are on the gastrointestinal tract leading to a general interference with digestion and absorption of components of the food and the observed negative nitrogen balance indicates that the

Table 6.2. *List of effects of dietary PHA on the digestive tract.*

(a) Binding to epithelial cells
   (i)   Damage to the brush border
   (ii) Endocytosis
   (iii) Interaction with brush border enzymes

(b) Interference with epithelial cell metabolism
   (i)   Hypertrophy and hyperplasia, polyamine-dependent growth
   (ii) Increased turnover

(c) Effects on gut endocrine cells
   Direct and indirect effects on systemic metabolism

(d) Interference with local (gut) immune system
   sIgA; IgE and systemic responses

(e) Modulation of microbial ecology of oral cavity and small intestine
   Increased adhesion and selective overgrowth

primary gut effects also have profound systemic consequences for the metabolism, growth and health of the animals.

Effects of PHA on the digestive tract in vivo

PHA has an extraordinarily high resistance to gut proteolysis *in vivo* (Pusztai, Grant & Palmer, 1975; Pusztai, Clarke & King, 1979b; King, Pusztai & Clarke, 1980a; Banwell *et al.*, 1983; Hara *et al.*, 1984). Indeed, when rats are fed on diets containing known amounts of highly purified PHA, up to 90% of the ingested lectin is recovered from the faeces in a form still fully reactive with rabbit anti-lectin antibodies and with its haemagglutination activity essentially intact (Pusztai, 1980). In contrast, other proteins in the diet are more efficiently digested and absorbed. Accordingly, as PHA remains intact and fully active during its passage through the entire digestive system, it exhibits the full spectrum of its biological activities.

(a)   *Binding to epithelial cells.* (i) Membranes of epithelial cells of the small intestine contain complex type oligosaccharide structures which are specifically recognized by PHA. As PHA is not degraded during its passage through the gut, it binds strongly to those gut epithelial cells which express such structures. The binding has been studied both in purified rat brush border membranes (Boldt & Banwell, 1985) and in isolated single cells from the rat small intestine (Donatucci, Liener & Gross, 1987). Small intestinal cells have numerous high affinity binding sites for purified PHA of the order of 12×10^6 per columnar cell with an association constant of about $15 \times 10^6 \, \text{M}^{-1}$ (Donatucci *et al.*, 1987). In addition, there are a great many binding sites of low affinity. However, these *in vitro* studies probably overestimate the number of binding sites available for PHA *in vivo* in the intact small intestine. Clearly, in cell cultures, PHA may bind not only to the microvilli but also to the lateral and basolateral membranes. The lectin in the lumen also combines with a number of glycoproteins and glycoprotein enzymes secreted by the surface cells (King *et al.*, 1980*a*). Profound changes ensue in cellular morphology, in metabolism of the small intestine and the proper functioning of the entire digestive system in consequence of the initial binding of PHA to the brush border epithelium (Pusztai & Greer, 1984; Greer, Brewer & Pusztai, 1985).

At the ultrastructural level, the initial binding of PHA to epithelial cells of the small intestine (Fig. 6.2) leads to extensive and severe damage to the brush border of the proximal small intestine (Fig. 6.2) of both rats (King, Pusztai & Clarke, 1980*b*; 1982; Hara, Tsukamoto & Miyoshi, 1983; Rossi, Mancini & Lajolo, 1984; Bulajic *et al.*, 1986; Lafont *et al.*, 1988) and pigs (King, Begbie & Cadenhead, 1983; Begbie & King, 1985). Moreover, both immunofluorescence (King *et al.*, 1980*a*) and immuno-gold (King *et al.*, 1986) localization studies have confirmed that the avidity of PHA binding runs parallel with the extent of the disorganization of the brush border. Initial studies have indicated that, as a result of PHA-binding, the height of small intestinal villi are slightly reduced, with the consequent effects of a reduction in the absorptive surface area and a decrease in the rate of absorption of amino acids, small peptides and other nutrients (King *et al.*, 1980*b*). However, further and more recent critical re-investigation has indicated that there is a small significant increase in villus height on exposure to dietary PHA (Pusztai *et al.*, 1988*b*). The slight enlargement of the absorptive villi is probably due to the ready supply of a substantially increased number of cells generated in the PHA-stimulated crypts.

The initial binding of PHA to surface membranes of the gut wall leads to severe disorganization and destruction of the normal morphology of the regularly folded structure of the microvillus membrane (Fig. 6.3) of the absorptive enterocytes of jejunal villi (King *et al.*, 1982). Moreover, probably as a result of the destruction of microvillus membranes, the normal rate of extrusion of villus tip cells is speeded up considerably and consequently the turnover of the small intestinal absorptive cells is also increased appreciably on PHA treatment.

(ii) A closer inspection of sections of the small intestine of rats exposed to PHA, even at the relatively low level of resolution of immunofluor-

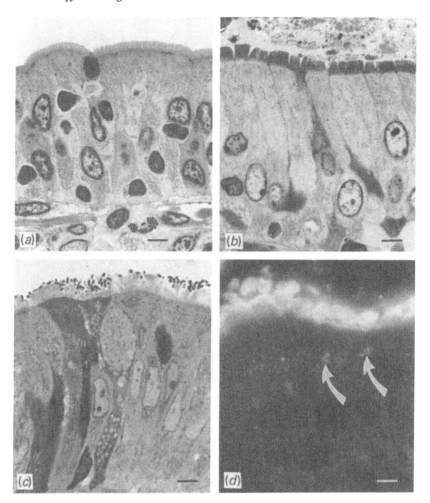

Fig. 6.2. (*a*) Light micrograph of part of a section through the jejunum of a pig fed on a non-toxic control diet, showing the undisrupted brush border. (*b*) Light micrograph of part of a section through the jejunum of a pig fed for 27 days on a diet containing raw kidney beans (cv. 'Processor'), showing the disrupted brush border. (*c*) Light micrograph of part of a section through the duodenum of a rat fed for 3 days on a diet containing 10% raw kidney bean (cv. 'Processor') protein, showing bacteria associated with disrupted brush border. (*d*) Immunofluorescence micrograph of part of a section through the jejunum of a pig fed for 27 days on a diet containing raw kidney beans (cv. 'Processor'), showing the presence of lectin on the damaged brush border and also within vesicles in the apical cytoplasm of the enterocytes (arrows). (*a*)–(*d*), scale bars = 25 μm.

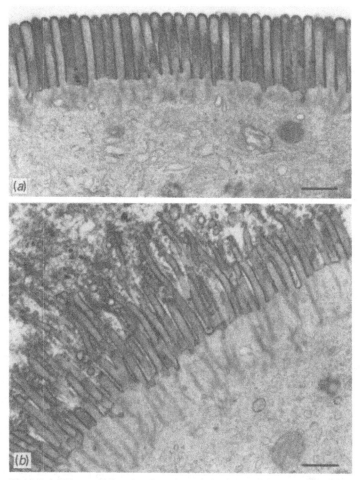

Fig. 6.3. (a) Transmission electron micrograph of part of the mucosal surface of jejunal enterocyte from a rat fed on a non-toxic control diet, showing the regularly arranged microvilli. (b) Transmission electron micrograph of part of the mucosal surface of a jejunal enterocyte from a rat fed for 3 days on a diet containing 10% raw kidney bean (cv. 'Processor') protein, showing disruption of the microvilli. (a)–(b), scale bars = 500 nm.

escence staining and light microscopy (Fig. 6.2(*d*)), has revealed that the lectin is not only bound to the brush border surface but also that a considerable proportion has penetrated into the interior of the surface cells (King *et al.*, 1980*a*). Furthermore, the presence of PHA-immunoreactive material in membrane-bound cytoplasmic subcellular particles (Fig. 6.4) is clearly visible in immunogold stained sections of the PHA-exposed rat small intestine at the electronmicroscope level. Thus, an appreciable part of the ingested PHA is localized in the cytoplasm of

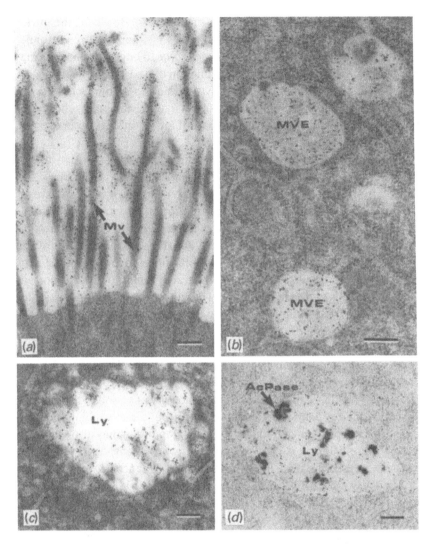

Fig. 6.4. Protein A-gold localization of dietary kidney bean (cv. 'Processor') lectins in the rat small intestine, showing characteristic appearance of jejunal microvilli and lysosomes. (*a*) Immunolabelling of elongated microvilli (Mv). (*b*) Immunolabelling of multi-vesicular endosomes (MVE) in the apical cytoplasm. (*c*) Immunolabelling of a lysosome (Ly) in the mid-cytoplasmic region of an enterocyte. (*d*) Deeper within the cytoplasm of enterocytes dietary lectins are sequestered within multi-vesicular bodies and other lysosomes (Ly). This micrograph, taken of a specimen which had been treated for the cytochemical demonstration of acid phosphatase (AcPase); identifies both enzyme and lectin in the same lysosome. (*a*)–(*d*), scale bars = 250 nm.

both enterocytes and goblet cells after endocytosis. However, the intracellular lectin is not free but is contained in either small endosomic
vesicles or in large endosomes, possibly *en route* to large lysosomes (King
et al., 1986). Despite the degradative function of lysosomes, the presence
of appreciable amounts of the endocytosed and immunoreactive lectin
can be clearly seen by immunogold staining in these catabolic subcellular
organelles suggesting that, at least, a part of the lectin is resistant even to
lysosomal degradation.

High rates of endocytosis by small intestinal cells appear to be
characteristic for lectins of all types, regardless of whether they are
derived from plant foods or bacterial pili. Accordingly, if this endocytosis
is a consequence of lectin function, the amounts of the lectin endocytosed
are expected to be drastically reduced in the presence of large amounts of
haptenic sugars of the correct specificity for the particular lectin tested.
This is precisely what occurs, giving further support to the suggestion that
proteins which possess lectin-like activities are preferentially transported
through the cells of the small intestine (Pusztai *et al.*, 1979*b*, 1981*b*; Grant
et al., 1985; de Aizpurua & Russel-Jones, 1988; Pusztai, 1989*c*; Pusztai,
Greer & Grant, 1989*b*).

(iii) Lectins in general and PHA, in particular, interact with brush
border enzymes and, from a nutritional point of view, the most important
of these is with the hydrolytic enzymes of the gut. Thus, both PHA and a
related enterokinase inhibitor inhibit the activity of various small intestinal endo- and exopeptidases and may, therefore, directly inhibit the final
stages of food protein degradation. Such an inhibition may also seriously
impair the efficiency of peptide and amino acid transport through the
epithelial cells of the small intestine (Rouanet, Besancon & Lafont, 1983;
Triadou & Audran, 1983; Tajiri, Lee & Lebenthal, 1986; Erikson *et al.*,
1985). However, interactions between lectins and brush border enzymes
are by no means confined to peptidases and they may also include the
binding of lectins to other enzymes, such as for example the reaction of
PHA with brush border invertase and ATP-ase (Rouanet & Besancon,
1979). Moreover, when PHA interacts with brush border hydrolases, the
loss of enzyme activity is not just due to the inhibition by the lectin of
enzyme activity but also to the actual loss of enzyme contained in small
vesicles from the microvilli. However, since different brush border
enzymes are lost to different extents, in addition to the well-known
process of membrane blebbing (Rouanet *et al.*, 1988), PHA must interfere with the synthesis of membrane hydrolases and/or may also cause
other forms of membrane erosion. It is certain that both direct and
indirect interferences with brush border hydrolases make sizeable contri-

butions to the overall detrimental effects of PHA on nutrient absorption. These effects are probably included in the reduced rates of absorption of all types of nutrients found in vascular perfusion experiments (Donatucci *et al.*, 1987) or *in vivo* feeding experiments (Pusztai *et al.*, 1981*b*).

(b)   *Interference with epithelial cell metabolism.* (i) One of the earliest events which occurs after the transport of PHA into the surface cells of the small intestinal epithelium is an immediate and highly significant stimulation by the lectin of the protein synthetic apparatus of the mucosal cells. A single dose of 10 mg of pure PHA given intragastrically, causes an increase of about 80% in the *in vivo* fractional rate of protein synthesis within one hour of the administration of the lectin (Palmer *et al.*, 1987). This effect observed with PHA is quite the opposite to the well-known cytotoxicity of highly toxic lectins, such as ricin or abrin. These are also endocytosed by eukaryotic cells, which is followed by their reaction with the 60S ribosomal subunits leading to the irreversible inhibition of cellular protein synthesis (Olsnes, Refsnes & Pihl, 1974). Despite the contrast between the effects of PHA and cytotoxic, 60 S ribosomal subunit-inactivating lectins, there are also some similarities in the reactions of the two types of lectins with epithelial cells of the small intestine (Keenan *et al.*, 1986).

It appears that the increased rate of protein synthesis in the small intestine after a first, short exposure to PHA is one of the early indications of a generally increased cellular metabolic activity. Indeed, because in the method of protein synthesis measurement employed, synthesis rates are measured 10 min after the injection of a flooding dose of ^{3}H-phenylalanine, the measured rates reflect only the synthesis of intracellular proteins and glycoproteins (Garlick, McNurlan & Preedy, 1980). As phenylalanine is a relatively minor component in the polypeptide backbone of the secreted mucinous glycoproteins of small intestinal goblet cells, the synthesis rates of mucins may not be included in the overall protein synthesis rates measured by this method.

PHA stimulates protein synthesis in small intestinal cells and behaves in a way similar to that found when the lectin induces the mitosis of peripheral lymphocytes *in vitro*. Indeed, PHA acts as an *in vivo* growth factor for brush border cells. The weight of the small intestine of young growing rats ingesting about 0.3 to 0.4 g of raw kidney bean proteins per day (equivalent to about 30 to 60 mg of pure PHA) is doubled after about 7 days on such diets. This weight change is not the result of accumulation of water in the tissue as the dry weight of the small intestine is also doubled. Moreover, the increase in the weight of the tissue is fully accounted for by its increased contents of DNA, RNA, proteins and carbohydrates (Greer *et al.*, 1985; de Oliveira *et al.*, 1988; Pusztai *et al.*, 1988*a,b*). There is a clear correlation between the amounts of kidney bean proteins/lectin fed and the extent of the small intestinal enlargement obtained. Furthermore, the growth of the small intestine persists for the entire duration of exposure to PHA and its weight is trebled in two weeks and nearly quadrupled after 3 weeks on such diets. The enlargement is due to both increased rates of mucosal cell proliferation (hyperplasia) and elevated levels of secretion of acidic mucinous glycoproteins (hypertrophy) by the small intestine (Fig. 6.5(*a*)).

All these changes in the morphology and metabolism of the small intestinal tissue are essentially quantitatively reproduced when the rats are fed on egg albumin-containing semisynthetic diets into which appropriate amounts of pure PHA have been incorporated (Fig. 6.5(*b*)) so as to match the lectin content of the original kidney bean diets used in previous experiments (de Oliveira *et al.*, 1988).

Chemical analyses have shown that, as the small intestine enlarges, its contents of DNA, RNA and proteins increase significantly in parallel with its growth (Table 6.2). However, the protein/DNA ratio has not changed, suggesting that growth has been mainly by hyperplasia and not by hypertrophy (Pusztai *et al.*, 1989*b*). Despite the considerable loss of bodyweight after feeding rats on a PHA diet for 3 days, the changes in the weight and composition of the small intestinal tissue have indicated that cell proliferation in the mucosa has increased considerably as a part of the PHA-induced high metabolic activity. Indeed, in a relatively short time after exposure to PHA, the morphological features of the small intestine have become characteristic of a rapidly proliferating tissue (Fig. 6.6). The crypts of Lieberkuhn, the proliferative compartment of the gut, have more than doubled in length within 3 days and the number of crypt cells has also doubled. In addition, crypt cell proliferation has been accompanied by an increase of about 25% in their size (hypertrophy). Although all cells have been counted, proliferation has been confined mainly to enterocytes and the number of goblet cells, Paneth cells or enterochromaffin type cells has been much less affected. As a result of the increased supply of cells from the crypts, the absorptive villi have also lengthened slightly. Overall, these changes in morphological features of the small intestine are in accord with biochemical and compositional data. It is particularly relevant that the villus/crypt ratio of about 4.5 found in normally fed, control rats has decreased significantly to a value of about 2.3 after exposure to PHA, which is an expected feature of the hyperplastic growth of mucosal tissue.

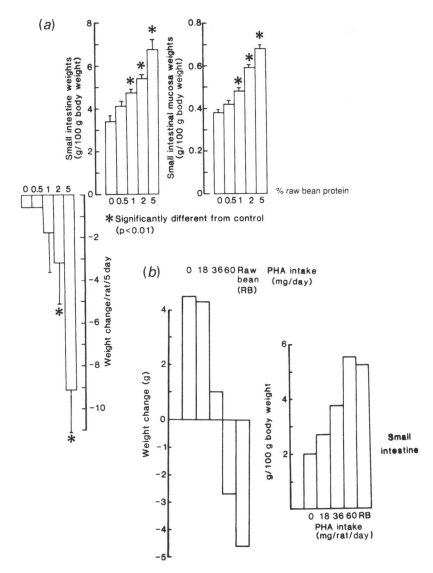

Fig. 6.5. Dependence of growth of rats and their small intestine on the amounts of (a) kidney bean proteins or (b) the pure lectin, PHA, in the diet. The results were obtained after feeding rats for 5 days on the respective diets.

Feeding rats on diets containing a different lectin preparation from red kidney bean which also contained an enterokinase inhibitor has caused hyperplastic growth in the rat small intestine (Tajiri *et al.*, 1988). In fact, most of the effects observed in this study are similar to the hyperplastic growth described previously with pure PHA preparations (de Oliveira *et al.*, 1988). The lectin-enterokinase inhibitor preparation has increased DNA synthesis and cellular proliferation in the crypts of the small intestine. It is, however, not very clear what contribution the enterokinase inhibitor contained in this somewhat impure lectin preparation has made to the induction of cellular proliferation (Tajiri *et al.*, 1988).

Studies into the biochemical mechanism of the growth of the small intestine have revealed that the PHA-induced growth is coincident with an increase in the polyamine content of the tissue (Pusztai *et al.*, 1988*a,b*). This is similar to the effects of other physiological stimuli, such as lactation (Yang, Baylin & Luk, 1984), weaning (Luk, Marton & Baylin, 1980), partial resection (Luk & Baylin, 1983), obstruction (Seidel, Haddox & Johnson, 1984) or simple re-feeding (Tabata & Johnson, 1986). The polyamine content of the jejunum obtained from rats fed on PHA-containing diets for various periods of time has continuously increased throughout. Moreover, the accumulation of polyamines and

(a)   (b)

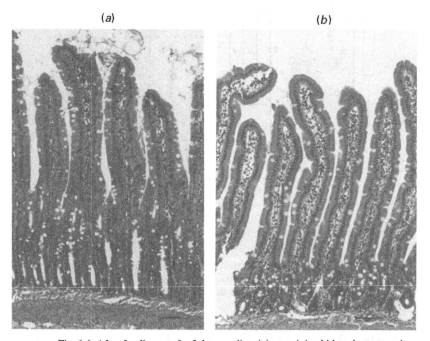

Fig. 6.6. After feeding rats for 3 days on diets (a) containing kidney bean proteins induces enlargement of the crypts of Lieberkühn in comparison with (b) lactalbumin-fed controls. (From Pusztai *et al.*, 1988 by permission of publisher.) Scale bar represents 50 μm.

Table 6.3. *Effect of PHA-stimulation on the weight, chemical composition and enzyme activities of ODC and SAT of 20 cm rat jejunum.*

Diet	Control (0 h)	PHA diet		
		1 day	3 days	7 days
Dry weight (mg)	135 ± 8^a	146 ± 6^a	214 ± 11^b	224 ± 37^b
RNA (mg)	7.5 ± 0.3^a	11.7 ± 0.6^b	32.3 ± 0.8^c	21.9 ± 3.0^d
DNA (mg)	5.4 ± 0.8^a	5.9 ± 0.3^a	8.6 ± 0.5^b	8.0 ± 0.6^b
Putrescine (nmol)	78 ± 15^a	24 ± 6^b	123 ± 23^c	186 ± 31^d
Spermidine (nmol)	452 ± 96^a	838 ± 42^b	1201 ± 123^c	1218 ± 153^c
Spermine (nmol)	354 ± 42^a	242 ± 40^b	435 ± 52^a	618 ± 78^c
Spermidine/Spermine	1.3	3.5	2.8	2.0
ODC (nmol/h)	2.0 ± 1.3^a	12.2 ± 9.4^a	5.8 ± 3.0^a	6.3 ± 3.0^a
SAT (nmol/h)	106 ± 55^a	136 ± 27	n.d.	120 ± 42^a

Note: Rats (groups of five) were fed on control (LA) or PHA diets for various lengths of time. Results are means $\pm$ SD. Values in a row with different superscripts (a,b,c,d) are significantly different (p < 0.001; Students' test; n.d. = not determined). ODC = ornithine decarboxylase; SAT = spermidine/spermine ^1N-acetylase.

particularly spermidine, has preceded the increase in the weight and protein and DNA contents of jejunal tissue (Table 6.2). The spermidine content of the jejunum has doubled within 24 hours without significant changes in other chemical parameters, with the exception of a rise in RNA content of the tissue (Table 6.3). The amounts of spermidine and other polyamines have increased steeply and continuously up to the third day of feeding, after which the increase in total polyamines has slowed down. Although the spermidine content has remained essentially unchanged after 3 days, the amounts of spermine have increased further (Pusztai *et al.*, 1988*b*).

It is of considerable interest to find out the source of the polyamines for growth since the coincidence of the PHA-induced enlargement of the small intestine with increased polyamine content suggests that polyamines may be involved in the biochemical mechanism of gut growth. Physiological stimuli, such as feeding after a period of starvation, lactation and resection increase the polyamine content of the tissue by stimulating the induction of ornithine decarboxylase (ODC), the enzyme responsible for *de novo* putrescine biosynthesis. As according to the polyamine interconversion pathway, with the participation of other enzymes of polyamine metabolism, such as *S*-adenosylmethionine decarboxylase (SAMDC), N'-spermine/spermidine acetylase (SAT) and polyamine oxidase, any individual polyamine can be converted to other polyamines, the stimulation of *de novo* putrescine biosynthesis may satisfy all physiological requirements for polyamines (Seiler, 1987).

During PHA-induced growth the stimulation of ODC activity has been slight (Table 6.3) and the amounts of putrescine synthesized by the small intestine cannot account for the large amounts of spermidine and other polyamines accumulated in the tissue. Thus, those polyamines required for growth which are not synthesized *in situ* may be derived from either the food or the peripheral organs and tissues. As the daily supply of polyamines from the food is slight (usually less than 0.2 mg, Bardocz *et al.*, 1990*a,b*) most are probably taken up by the small intestine from blood.

In view of the observation that polyamine accretion in the gut was not dependent on *de novo* biosynthesis via ODC, it was unexpected that α-difluoromethylornithine (DFMO), an irreversible inhibitor of ODC activity and polyamine biosynthesis caused a reduction in the PHA-induced rise of polyamines, DNA, RNA and protein contents of the small intestinal tissue. However, morphological studies have revealed that DFMO has not abolished crypt hyperplasia induced by luminal PHA, only reduced the extent of it (Fig. 6.7). Polyamine pools of the body are partially depleted on prolonged treatments with DFMO. As the polyamine supply from peripheral tissues diminishes, the differentiation and maturation of crypt cells may become defective in the PHA-stimulated small intestine and as a result cell death occurs (apoptotic bodies). Consequently, the non-viable cells never reach the absorptive villi but are removed from the crypts. Accordingly, the growth stimulation by PHA operates even in the presence of DFMO and cellular proliferation proceeds largely unimpeded. However, due to the reduced number of cells reaching the villi, villus atrophy occurs. This may account for the partial reversal of the PHA-induced growth and for the reduction in the accumulation of DNA, RNA and protein in the tissue when the kidney bean diets also contain DFMO (Bardocz *et al.*, 1989*a*).

Formation of apoptotic bodies in the jejunum of rats treated with DFMO for seven days in the presence of PHA is particularly striking. Although such a process can also be observed occasionally in DFMO-treated controls, it appears that the increasing demand for polyamines exacerbates the defect when growth of the small intestine is induced by treatment with PHA.

Apoptosis can also be induced with other lectins or even with 60S ribosomal subunit-inactivating toxins. However, the reaction mechanism

(a) 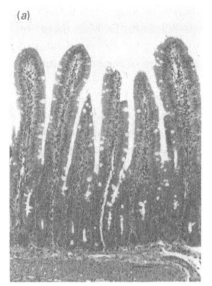 (b)

Fig. 6.7. In the presence of DFMO (α-difluoromethylornithine) villus atrophy occurs in both (*a*) kidney bean-fed and (*b*) control rat small intestine. DFMO reduces the extent of the PHA-induced hyperplastic growth, while the controls are slightly hyperplastic in comparison with controls without DFMO. (From Bardocz *et al.*, 1989a; by permission of the publisher.) Scale bar represents 50 μm.

of apoptosis with them may be different from that observed with the PHA-stimulated gut in the presence of DFMO. Despite this difference, the end result, i.e. the increased rate of crypt cell proliferation, may still be similar with both lectins and toxins. For example, the classic Shiga toxin derived from *Shigella dysenteriae* Type-1, Strain 60R, reacts with ileal enterocyte surface receptors. The gut cells absorb the toxin by endocytosis and transport it to phagolysosomes, where processing occurs. As a result, apoptosis of the fully differentiated and mature villus cells occurs by an active process of cell death. In order to replace the prematurely expelled cells from the villus, the proliferative compartment is activated and the production of new cells in the crypts is speeded up. This process fully accounts for the reduction in villus/crypt ratios found in the Shiga toxin exposed ileal tissues (Keenan *et al.*, 1986). As apoptosis is an active process, it initially requires an increased rate of protein synthesis (Willie, Kerr & Currie, 1980) and, in a rather curious way, one of the first effects of exposure of gut tissues to injurious toxins is a requirement for increased protein synthesis. Despite possible differences in the mechanism of apoptotic body formation by Shiga-type cytotoxins and the combined effects of non-cytotoxic PHA plus DFMO, there are also several similarities, including a requirement for increased rates of protein synthesis in the tissue and the stimulation of the crypt compartment.

As the polyamine requirement of the growth process of the small intestine stimulated by PHA cannot be satisfied by the *de novo* biosynthesis of polyamines in the tissue, the mechanism of polyamine uptake from either the luminal or basolateral side assumes great importance. The amounts of polyamines which may be taken up from the luminal side are small because the polyamine content of various foods is low. However, polyamines may also be synthesized by bacteria in the small intestine, particularly when infected, and these may also be available for absorption. Therefore, studies of the transport of polyamines from the lumen into small intestinal tissue may also give an insight into the process of cell proliferation in both the stimulated and the control guts.

Studies of the uptake of labelled polyamines have revealed that the wall of the small intestine is permeable to all three and that only 10% of the polyamines introduced intragastrically remains in small intestinal tissue. Most of the luminally administered polyamines leave the small intestine within the first hour and become distributed throughout the tissues of the body (Fig. 6.8(*a*),(*b*),(*c*)). The transport of the polyamines is dependent only on their concentration and the proportion delivered to the body remains unaltered by a 50-fold increase in the initial concentration of the polyamines in the small intestinal lumen (Bardocz *et al.*, 1990*a, b*).

There are no significant differences in the concentration coefficients for the uptake of the radioactive putrescine by the small intestine or by the body between the PHA-stimulated and control rats at any of the putrescine concentrations examined. However, about two-thirds of the labelled putrescine taken up by the small intestine is converted into non-polyamine type compounds during the first hour, and only about 11–15% of the original dose remains in the form of putrescine (Bardocz *et al.*, 1990*a,b*). This high rate of conversion of putrescine into non-polyamine type compounds does not apply to spermidine or spermine which are largely conserved in polyamine form.

The concentration coefficients of the uptake rates for spermidine and spermine are also rather similar in both PHA-stimulated and control gut tissues, but the amounts of spermine reaching the body through the PHA-damaged small intestine are slightly higher than those of the controls.

Thus, luminally administered polyamines and, presumably, those synthesized by bacteria in the gut are taken up by the small intestine and the body by passive diffusion. The amounts absorbed are linearly dependent on the polyamine concentration in a wide (0.1 to 5 mg) range.

One of the most important findings militating against luminal uptake as the major route of absorption to satisfy the increased demand of the growth process for polyamines is the convincing demonstration that stimulation of the small intestine by PHA does not increase the amounts absorbed from the lumen (Bardocz *et al.*, 1990*a,b*). Indeed, if anything, the small intestine stimulated to grow by PHA is, apparently, taking up slightly less polyamines than the controls. In the experiments with PHA, rats consume daily about 0.1–0.2 mg of polyamines. The maximum attainable accretion of individual polyamines by the small intestine under these conditions is about 30 nmol per day. This is clearly well below the much higher amounts of the about 500–600 nmol spermidine accumulated in the PHA-stimulated small intestine within the first 24 hours of exposure (Pusztai *et al.*, 1988*b*). Thus, it is unlikely that food can serve as a major source of polyamines for the growing gut (Bardocz *et al.*, 1990*a,b*).

Further evidence against the involvement of the luminal uptake route as a major component in the delivery of polyamines comes from experiments with DFMO. The polyamine pools of the small intestine and the body are partially depleted when DFMO is present in the food and drinking water, although the PHA-induced crypt cell proliferation rate is only slightly affected. With this limitation in the supply of polyamines, a sizeable proportion of the cells produced is not viable. The defect is probably due to the inability of a proportion of the newly produced cells to differentiate and mature fully at low intracellular polyamine concentrations. However, even giving relatively high polyamine supplements, equivalent to about 5 mg individual polyamines per rat, does not readjust the tissue polyamine concentrations fully to pre-DFMO levels, or restore the growth of the small intestine (Bardocz *et al.*, 1989*a*).

On the other hand, polyamine uptake through the basolateral membranes appears to respond to stimulation by luminal PHA (Pusztai *et al.*, 1989*b*; Bardocz *et al.*, 1990*a,b*). The amounts of intraperitoneally injected labelled polyamines absorbed by the small intestine accumulate in the tissue in a time-dependent way during the course of the stimulation of small intestinal growth by PHA (Table 6.4). Thus, the amount of putrescine taken up via the basolateral membranes is increased significantly within the first hour of the stimulation by PHA, whereas the

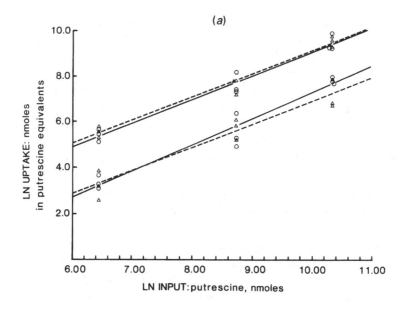

(a)

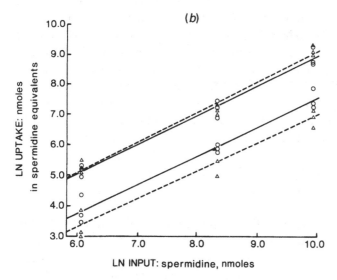

(b)

amounts of spermidine and spermine absorbed by the small intestine are only elevated appreciably 7–8 h after the treatment. Moreover, while the uptake of putrescine reaches its maximum after one day, the transport of spermidine or spermine into the small intestine through the basolateral membrane increases steadily during the entire seven day period of the

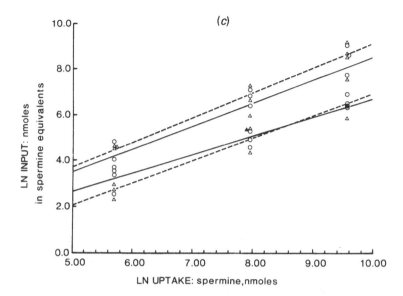

Fig. 6.8. The uptake of intragastrically administered (*a*) putrescine, (*b*) spermidine and (*c*) spermine (nmol ± SEM) by the body (upper lines) and the small intestine (lower lines). Rats (groups of 3) were fed on control (LA, lactalbumin) or PHA diets for 3 days, after which they were given different mixtures of unlabelled and labelled polyamines. The individual polyamines were tested at 10, 100 and 500 nmoles initial concentrations and the uptake is expressed as a proportion of the initial amounts given by intragastric intubation. Three experimental values (one for each rat) were obtained at each polyamine concentration. Circles represent controls and the triangles denote the results obtained with PHA-fed rats. The lines (solid for controls and the broken line for the PHA-treated) represent the best statistical fit (95% confidence limit) to the experimental points. (From Bardocz *et al.*, 1990*b*; by permission of *Digestion.*)

experiment. As blood polyamine concentrations are low, this accumulation from the blood supply is probably against a polyamine gradient (Bardocz *et al.*, 1989*b*).

The ready uptake of polyamines on growth stimulation by a transport system located on the basolateral side of the rat small intestine and the changes in the amounts absorbed by it suggest that this route of polyamine uptake may operate under most physiological conditions. Most importantly, the existence of such a system satisfies the main requirement for the PHA-induced growth, namely that this route is stimulated when PHA induces the growth of the small intestine. It appears that this basolateral route may be the main polyamine delivery system for the gut.

Although the increase in the dry weight of the small intestine is also time-dependent during the PHA-stimulated growth, the stimulation of

Table 6.4. *Time-dependence of the stimulation by PHA of small intestinal basolateral polyamine uptake, expressed as a percentage of that of the lactalbumin-fed control rats.*

Time (hours)	1–2	7–8	26–27	74–75	170–171
Dry weight (% of control)	100 ± 5	82 ± 4	108 ± 4	136 ± 5	152 ± 6
Polyamine uptake (% of control)					
Putrescine	128 ± 3*	161 ± 11*	194 ± 21*	202 ± 32*	202 ± 28*
Spermidine	113 ± 12	161 ± 34*	178 ± 28*	219 ± 21*	322 ± 43*
Spermine	98 ± 2	176 ± 33*	182 ± 33*	242 ± 36*	339 ± 26*

Note: Rats (group of 3) were fed on control (LA) or PHA diet for up to 7 days to stimulate the growth of their small intestine. After overnight fast, they were given 1.5 g of maize starch with or without 10 mg of pure PHA-E$_3$L lectin. One hour after feeding, the rats were injected i.p. with ^{14}C-polyamines and killed 1 hour later. The results are expressed for the absolute amounts of labelled polyamines accumulated in the small intestine of control rats in the first hour after injection. These values, nmoles/small intestine, were: putrescine, 0.85 ± 0.05; spermidine, 0.54 ± 0.07; and spermine, 0.59 ± 0.03. (From Bardocz *et al.*, 1990*b*; by permission of *Digestion*.)

* Significant differences between PHA-treated and control groups, p < 0.01 (Students' t test).

the polyamine transport system actually precedes growth, in keeping with the observation that polyamine accumulation in the small intestine also precedes growth (Pusztai *et al.*, 1988*b*). The demonstration of the intracellular presence of labelled polyamines by autoradiography has confirmed that those transported from the basolateral side of the small intestine are truly taken up by the cells and not just adhere non-specifically to the membranes. Although a major part of the labelled putrescine, 74–90% of the original dose, becomes converted into non-polyamine type compounds, most of the labelled spermidine or spermine is conserved in the tissue after absorption.

It appears that the basolateral transport system for the uptake of spermidine or spermine is not appreciably affected by DFMO (Bardocz *et al.*, 1990*a,b*). Although with continuous DFMO treatment the polyamine pools of the body are partially depleted, the capacity of the transport system is not reduced; only the amounts of polyamines available for transport are diminished. Consequently, the content of the small intestinal tissue is reduced in both PHA-treated and control rats, but, in the presence of PHA, the extent of the reduction in the polyamine content due to DFMO is less than that in the non-stimulated control rats. This supports previous findings that the PHA-induced crypt cell proliferation is only reduced by DFMO but not abolished entirely (Bardocz *et al.*, 1990*a,b*).

The results of the basolateral uptake experiments may also explain the often found lack of correlation between the distribution of polyamines and their metabolic enzymes in the different functional compartments of the small intestine (Luk & Yang, 1987, 1988). Growth stimulation usually results in the induction of ODC activity and *de novo* putrescine biosynthesis and therefore ODC induction in the tissues is often regarded as a necessary prerequisite for small intestinal growth. However, both these changes occur on the villus tips which contain fully differentiated cells, while both ODC activity and the concentration of putrescine in the crypts are low where the proliferation-based hyperplastic growth occurs. It is thus difficult to see how the polyamines synthesized on the villus tips can reach the crypt cells and satisfy their high spermidine/spermine requirements for cellular proliferation that occurs in this compartment. In contrast, as basolateral polyamine transport into the small intestine can be highly stimulated by PHA, crypt cells can take up polyamines from circulation as required by the extent of the growth stimulation. It is also highly significant that PHA-dependent growth is much more dependent on the accumulation of spermidine (or spermine to a lesser extent) than that of putrescine.

The dependence of small intestinal growth on polyamine uptake through the basolateral membranes may not apply only to PHA-induced growth. Several other stimuli which are independent of or only slightly dependent on ODC activation have also been reported to induce gut growth. For example, gastrin can increase growth without ODC induction (Seidel *et al.*, 1985). Although the synthesis of ODC is increased on gastrin-induced growth of the stomach and duodenal and colonic mucosa, this is slight in the crypts and the increase in ODC induced by other luminal factors occurs also on the villus tips but very little in the crypts (Johnson *et al.*, 1988, 1989). It appears that accumulation of polyamines in the small intestine from circulation may be adapted to changing conditions of growth requirements and stimuli more readily and flexibly than through ODC induction. However, regardless of the precise nature of the initial response by the small intestinal cells to the growth signal, it is clear that there is an obligatory accumulation of polyamines in small intestinal crypts after the binding of growth factors (PHA, gastrin, etc). This is followed by secondary responses which include increases in RNA, DNA and protein contents (Pusztai *et al.*, 1988*a*; Bardocz *et al.*, 1989*a*) as a part of the proliferative process.

The cells on the villus tips are fully differentiated. Their response to various luminal stimuli, including re-feeding, probably includes ODC induction (Johnson *et al.*, 1989). However, as the villus tip cells do not proliferate or differentiate, they may have a lower requirement for polyamines, which may well be satisfied by their limited *de novo* synthetic capacity for putrescine generation.

Increased concentrations of polyamines are also known to regulate the polymerization state of cellular cytoskeleton proteins including actin and tubulin (Rumsby & Puck, 1982; Grant, Oriol-Audit & Dickens, 1983; Anderson *et al.*, 1985). Accordingly, the disorganization of the microvillus membrane and underlying structural components by the initial PHA-binding to the membrane may be related to changes in cellular polyamine levels. The serious disruption of the terminal web structures in the epithelial cells also points to such an effect by PHA on cytoskeletal proteins (King *et al.*, 1982). PHA is not known to interact directly with either actin or tubulin or other cytoskeletal proteins and this suggests that the effect on them is indirect and probably mediated by changes in intracellular polyamine concentrations. Indeed, the polymerization state of cytoskeletal actin in the Caco-2 cells is affected by soya lectin. A decrease in intracellular F-actin occurs within 15 min of exposure to the agglutinin. As the presence of *N*-acetyl-D-galactosamine prevents the depolymerization of actin, this effect is clearly the result of the lectin activity of soyabean agglutinin (Draaijer *et al.*, 1989).

PHA does not only interact with enterocytes of the absorptive epithelium, but also binds to goblet cells. This binding is also followed by endocytosis (King *et al.*, 1986). Furthermore, there is a clear correlation between the amounts of acidic mucinous glycoproteins secreted in the gut and the amounts of PHA/kidney bean proteins in the diet (Greer *et al.*, 1985; Pusztai, 1989*a*). This is a direct effect of the lectin (Pusztai, 1989*a*), although other kidney bean seed proteins may also cause overproduction of endogenous mucin-type glycoproteins in the alimentary tract by an unrelated mechanism (Santoro, Grant & Pusztai, 1988). However, it is not clear how this measurable PHA-induced endogenous protein/mucin secretion is related to the vaguely defined 'mucotractive' effects of lectins (Freed, 1979; Freed & Buckley, 1979). All the same, by whichever mechanism it occurs, increased losses of endogenous nitrogen-containing components from the body, mainly in the form of mucinous glycoproteins secreted by alimentary hypertrophic goblet cells, have serious detrimental effects on the overall nitrogen-balance of the body but the contribution of this protein loss to the overall nutritional toxicity of PHA is not known. The effect appears to be a general response by gut goblet cells to the exposure to injurious toxins since the more highly toxic bacterial lectins, such as the cholera toxin, can also stimulate the secretion of mucins.

(ii) Although it is customary to consider cell turnover and proliferation in the small intestine separately, in practice both are but a part of the multitude of its responses to changing conditions and stimuli in the gut. Indeed, it is difficult to decide which of these two processes drives the other. For example, the indirect damaging effects of PHA on the cytoskeleton of microvilli and terminal web structures are probably polyamine-mediated and they may speed up cell maturation and senescence and lead to their premature removal from small intestinal villi. There is a net increase in cell turnover as cell replacement accelerates to maintain the size of the absorptive villi. Other injurious toxins of bacterial or food origin may similarly disrupt the highly ordered structure of the epithelial membranes by binding to brush border membrane receptors and in the process, cells with defective membranes may become non-viable. These toxin-damaged cells are removed prematurely from the absorptive villi and their place is taken by fresh cells coming up from the crypts. In this sense, cellular proliferation needs to increase to make good the initial damage by the toxin and this leads to increased turnover. Apparently, the Shiga toxin-induced villus cell apoptosis may be one of the prime examples of this type of stimulation of increased turnover (Keenan *et al.*, 1986).

The lectin, PHA, from kidney bean probably provides one of the best examples of the induction of increased turnover through a direct stimulation of the proliferative compartment of the small intestine. Due to the increased supply of cells from the crypts, the length of the villi expands slightly (Pusztai *et al.*, 1988*b*) and because of the increasing pressure of the cells migrating up the villus this is followed by an increased rate of premature removal of villus tip cells. As shown by immunofluorescence staining using antibodies raised against crypt cells (King, unpublished observations), the absorptive villi may contain a significant proportion of not fully mature cells on sustained stimulation of intestinal growth by PHA.

Increased turnover is a multifactorial process. It is the result of interference in cellular metabolism and structure by the initial binding of factors, such as PHA, bacterial and other toxins. The various events and reactions are probably concurrent and may even be synergistic. However, the physiological and nutritional consequences of the increased rate of cell proliferation and turnover are potentially serious for the animal whichever mechanism is operating in the small intestine. Although such a defence mechanism may have developed during evolution as the most efficient way of reducing the danger of transfer of potentially poisonous substances and toxins into the body through a compromised gut wall, the animal may have to pay serious nutritional penalties. The protein and energy cost of maintaining high rates of protein, DNA and RNA syntheses for supporting proliferation and turnover is high (Table 6.5). For example, practically the entire dietary intake must be used for the maintenance of protein synthesis in epithelial cells in the second week of feeding rats on PHA-containing diets. Accordingly, although these animals eat food, because they have to maintain the integrity of the intestinal wall, the body's needs come second. Thus, as food is not being utilized for the direct benefit of systemic metabolism and the body, ultimately, the toxicity of PHA for the animal is not confined to the damage to the small intestine or to the nutritional and physiological problems associated with the stimulation of gut growth.

(c)   *Effects on gut endocrine cells – direct and indirect effects on systemic metabolism.* PHA binds to all cell types of the small intestinal epithelium. However, interactions between PHA and gut endocrine cells have not been studied in detail so far, although a definite binding of the lectin to endocrine cells has been shown to occur by immunofluorescence staining with monospecific antilectin antibodies (King, Pusztai & Grant, unpublished observations). Unfortunately, the effects of this binding on the function of the endocrine cells are not known. It appears from preliminary observations that endocrine cells are not involved in the cellular proliferation induced by exposure to PHA in contrast to enterocytes of the small intestinal epithelium (Pusztai *et al.*, 1988*b*).

Apart from the actual observation of lectin binding to endocrine cells of the small intestine, most of the experimental evidence implicating them in the toxic effects of PHA is circumstantial and indirect. Nevertheless, definite effects of luminal lectins on endocrine cells in the small intestine may be inferred from the very pronounced and serious consequences of ingestion of PHA-containing diets on general systemic metabolism. Some of these effects resulting from the luminal exposure of animals to PHA may be interpreted to suggest that binding of the lectin to gut endocrine cells can release gastrointestinal hormones into systemic circulation and these hormones directly influence the metabolic pathways of the body. Alternatively, PHA-binding to endocrine cells of the alimentary tract can indirectly modulate the systemic endocrine hormone balance of the body. The ensuing changes in metabolic rates are,

Table 6.5. Approximate protein cost of PHA-induced hyperplasia and brush border damage in rat small intestine.

Diet	Small intestine weight/body weight	Small intestine protein content (mg)	Approximate cell turnover time (h)	Daily protein requirement (no recycling) (mg)	% of daily protein intake (600–700 mg)
Control	0.02	250–300	36–48	120–150	>20
Kidney bean (5%, 7 days)	0.04	500–600	24	500–600	>60–90
Kidney bean (5%, 14 days)	0.06	750–900	>24	700–900	>>100

therefore, the result of changes in the concentrations of endocrine hormones. However, it must be pointed out that some of these systemic effects may also be adequately explained by the already described hormone-mimicking effects of PHA (pp. 105–7). This type of modulation of systemic hormone levels is suggested by the observation that the rate of endocytosis of most lectins, and particularly that of PHA, by small intestinal cells appreciably exceeds that of other food proteins and that a proportion of the endocytosed lectin finds its way into the systemic circulation (Pusztai, 1989c; Pusztai, Greer & Grant, 1989b). It is not known at present which mechanism is operative and indeed, it is possible that all three mechanisms may operate simultaneously. As it is likely that all systemic effects are the results of the initial PHA-binding to epithelial cells of all types, including endocrine cells, the description of such systemic consequences of the dietary exposure to PHA appears to be appropriate under this heading.

The finding of highly elevated concentrations of urinary urea is one of the most striking effects observed in rats which have been fed on diets containing PHA, indicating an accelerated rate of tissue catabolism (Pusztai et al., 1981b). There is also a direct relationship between the reduction of the rate of growth and the increased urinary loss of urea nitrogen and both these effects result from the inclusion of kidney bean proteins in the diet. By incorporating pure PHA in nutritionally excellent semi-synthetic diets, the urea nitrogen loss observed with kidney bean feeding can be quantitatively reproduced. Thus, there is a clear correlation between the extent of urinary urea excretion and the amounts of pure PHA reaching the small intestine (Pusztai, 1989a). For biological activity the lectin must be intact because substitution of native PHA in the diet with a denatured preparation abolishes the catabolic effects on protein metabolism in PHA-fed rats.

Luminal exposure to PHA also increases body fat catabolism and glycogen loss (Grant et al., 1987a). Thus, consumption of kidney bean proteins leads to depletion of the lipid stores of the body. As the depleted depot lipids in subcutaneous tissues are initially replaced by water, the loss of lipid becomes more clear-cut after drying the body. In fact, the removal of stored fat from the body proceeds faster than the breakdown of tissue protein and this rate difference leads to an increase in the relative nitrogen concentration of the body although the absolute amounts of body protein are decreasing continuously during PHA exposure (Grant et al., 1987a; Pusztai, 1989a). The extent of this increase in the body's nitrogen concentration is dependent on the amounts of pure PHA in the diet (Pusztai, 1989a).

Compared with that of pair-fed controls, the near fivefold increase in the urinary output of 3-hydroxybutyrate obtained with 1% pure PHA in the diet has demonstrated that the depletion of depot lipids is, to a large extent, due to true lipid catabolism and not to the poor absorption of dietary lipids. In keeping with this, there is only a slight effect on the digestibility and absorption of lipids from the diet, even in the presence of high concentrations of luminal PHA (Pusztai, 1989*b*). Another finding, which suggests that PHA may increase fat catabolism rather than reduce its absorption from the diet, is that the concentration of free fatty acids in the serum increases steadily during a 10 day exposure to kidney bean proteins without changes in lipid absorption by the small intestine in the corresponding experimental period. Furthermore, the body lipid concentration of rats fed on protein-free diets is maintained or undergoes a slight but significant increase. Accordingly, the loss of lipids must be a direct effect of the lectin and not due to increased energy expenditure to spare body protein. Most of the lipids lost consist of triglycerides from subcutaneous tissues (de Oliveira, 1986) while the lipid concentration of most internal organs, such as the liver and kidneys, remains approximately the same. Similarly, the structural phospholipid concentration in the body is relatively little affected in the PHA-fed rats.

The glycogen content of the body is seriously depleted during exposure to dietary PHA. The total glycogen content, which is in the range of 0.45 to 0.48% of body weight in rats pair-fed on control diet, or even protein-free diets, falls to about 0.30% on kidney bean protein-feeding after 10 days (Grant *et al.*, 1987*a*). Increased glycogen catabolism is another systemic effect of PHA, although the breakdown of the polysaccharide may not necessarily occur to spare body protein. It is also clear that, in contrast to liver glycogen, its concentration in skeletal muscles is not significantly affected even after 10 days on PHA-based diets while, during the same period, the glycogen content of the liver is halved (Pusztai, 1989*a*).

The average serum concentration of glucose changes very little despite the serious losses of all body constituents, proteins, lipids and glycogen caused by exposure to PHA. Even after 22 days on raw kidney bean-containing diets, the average serum glucose concentration only falls from 1.8 mg glucose per ml of serum in pair-fed egg albumin-containing controls to 1.4 mg/ml in bean-fed rats (Grant *et al.*, 1987*a*). Despite the relatively slight changes in serum glucose levels in PHA-fed rats, the exhausted state of the energy reserves of the body in these animals can be gauged from the drastic changes in serum glucose concentration which occurs when the rats fed on kidney bean proteins for 22 days are fasted overnight. The serum glucose concentration in these rats falls dramatically to about 0.6–0.7 mg/ml, while all other groups including the protein-free control group, experience only a relatively slight drop in the blood sugar level. Glycaemic homeostasis is maintained mainly by glycogenolysis during the first 8 to 12 hours of fasting. However, the glycogen content of the liver in rats fed on PHA-containing diets is less than half of that of controls. Thus, the already depleted liver glycogen stores of PHA-fed rats are quickly exhausted on fasting. Rats fasted overnight without the intake of dietary carbohydrates must rely on generating glucose via gluconeogenesis. However, because of the highly elevated rates of catabolism of all body components during extended PHA-feeding, the gluconeogenesis pathways are also deficient. Since glycerol and amino acids necessary for glucose generation are also in short supply, bean-fed rats cannot maintain a physiologically normal blood sugar concentration when deprived of nutrients for even relatively short periods of time (such as overnight). The results also demonstrate, albeit indirectly, that as serum glucose levels are maintained in the normal concentration range during the entire 22 day feeding period, appreciable amounts of glucose must be absorbed from the daily food via the small intestine despite the serious damage caused to the brush border by the extended exposure to PHA.

Exposure of the lumen of the small intestine to kidney bean proteins containing PHA or purified preparations of PHA affects the weight, tissue composition and physiological function of a number of key organs or tissues of the body in addition to the modulation of the general systemic metabolism of animals (Table 6.6; de Oliveira *et al.*, 1988). Quite probably, these systemic effects of dietary PHA are also the result of the interactions between the gut and the systemic endocrine cells induced by the initial PHA-binding to small intestinal epithelial cells. Inclusion of pure PHA in egg albumin-based diets induces a dose-dependent and highly significant enlargement of the pancreas and a smaller but significant increase in liver weight in addition to the well-described growth of the small intestine (de Oliveira *et al.*, 1988). In contrast, PHA causes thymus atrophy and a significant loss of weight of the hind leg muscles, plantaris and gastrocnemius. This loss becomes significant as early as after 2–4 days of dietary exposure to the lectin (Palmer *et al.*, 1987) but none of the other organs of the body shows significant changes.

The dose-dependent enlargement of the liver induced by PHA is significant but slight and the underlying reaction mechanism is not clear. Since the concentration of lipids is unchanged, while the glycogen content

Table 6.6. *Dry weight of tissues and organs of rats pair-fed with diets containing different amounts of pure kidney bean lectin or raw kidney bean.*

Diet	Lectin 0 g/kg diet (n = 4)	Lectin 3 g/kg diet (n = 3)	Lectin 6 g/kg diet (n = 3)	Lectin 10 g/kg diet (n = 3)	Raw bean protein 50 g/kg diet (n = 4)	Denatured lectin 10 g/kg diet (n = 2)
Lectin intake (mg/rat/day)	0	18	36	60	50–55	60
Body weight (g) (Initial)	80.9 ± 2.3[a]	82.8 ± 3.0[a]	83.2 ± 2.5[a]	84.0 ± 1.9[a]	84.2 ± 2.2[a]	82.4 ± 2.1[a]
Body weight (g) (Final)	87.6 ± 1.3[a]	86.7 ± 2.3[a]	86.3 ± 5.1[a]	79.9 ± 1.6[b]	78.5 ± 0.3[b]	88.6 ± 0.8[a]
Dry body weight (g) (Final)	25.6 ± 0.2[a]	24.6 ± 1.1[a]	24.4 ± 1.7[a]	21.7 ± 0.3[b]	19.9 ± 1.1[c]	25.8 ± 0.2[a]
Tissue weights (g/100 g body weight)						
Small intestine	2.03 ± 0.19[a]	2.77 ± 0.61[b]	3.72 ± 0.49[c]	5.51 ± 0.12[d]	5.25 ± 0.30[d]	2.03 ± 0.12[a]
Pancreas	0.34 ± 0.04[a]	0.40 ± 0.01[b]	0.47 ± 0.11[b]	0.54 ± 0.10[c]	0.56 ± 0.07[c]	0.39 ± 0.05[a]
Liver	3.32 ± 0.05[a]	3.37 ± 0.08[a]	3.71 ± 0.24[b]	4.36 ± 0.35[c]	4.35 ± 0.07[c]	3.33 ± 0.10[a]
Thymus	0.20 ± 0.02[a]	0.15 ± 0.03[b]	0.16 ± 0.03[b]	0.09 ± 0.02[c]	0.05 ± 0.02[d]	0.20 ± 0.01[a]
Soleus*	0.12 ± 0.01[a]	0.11 ± 0.01[a]	0.11 ± 0.01[a]	0.10 ± 0.01[a]	0.10 ± 0.01[a]	0.11 ± 0.01[a]
Plantaris*	0.18 ± 0.01[a]	0.18 ± 0.01[a]	0.16 ± 0.01[a]	0.14 ± 0.01[b]	0.12 ± 0.01[b]	0.17 ± 0.01[a]
Gastrocnemius*	1.00[a]	0.90[b]	0.88[b]	0.73[c]	0.65[c]	0.96[a]

Note: Rats were pair-fed (6 g of diet/rat/day) on diets for 10 days whose overall protein content was 10% w/w. In test diets a part of the standard protein, egg albumin, was substituted with either pure PHA or with raw bean as indicated. The results are means ± SD and values in a horizontal row with different superscripts (a,b,c,d) differ at least to a level of 95% confidence. * Muscle weights are given per pair of hind leg muscles.

Source: From de Oliveira *et al.*, 1988, reproduced by permission of *Nutrition Research.*

is depleted, a major part of the enlargement of the liver is probably the result of protein accretion in the organ. As there are no significant changes in cell content of the tissue, the growth is probably by hypertrophy. The physiological implications of the slight liver hypertrophy are not known at present (de Oliveira *et al.*, 1988).

In contrast, after 10 days of dietary exposure PHA causes a severe atrophy of the thymus, first described in rats fed on kidney bean protein-containing diets (Greer *et al.*, 1985). However, experiments done with pure PHA have demonstrated unequivocally that the atrophy is due to the lectin content of kidney bean diets (Table 6.6; de Oliveira *et al.*, 1988). The consequences of the PHA-induced thymus atrophy may be very important for the growth and health of the animals since this organ plays a key role in the proper functioning of the immune system. It may also be significant that rats made diabetic with streptozotocin behave, at least superficially, in a way similar to PHA-treated rats. In both groups of rats the small intestine enlarges and the thymus atrophies (Chatamara *et al.*, 1985). There also appears to be a common hormonal mediation in these two effects, i.e. the concentration of circulating immunoreactive insulin is low under both conditions (Grant *et al.*, 1987*a*).

One of the most important effects of the exposure of rats to dietary PHA is the highly significant and dose-dependent pancreatic enlargement after a few days. Up till recently only dietary trypsin inhibitors have been known to induce pancreatic growth and it is widely assumed that this enlargement is hormonally mediated. Thus, according to current hypotheses, a hormone, cholecystokinine (CCK) is released into the circulation after the dietary trypsin inhibitors react with trypsin in the lumen of the proximal small intestine. One of the effects of this hormone is the stimulation of the exocrine pancreas which results in increased secretion of trypsin into the lumen of the small intestine. Thus, trypsin inhibitors induce the hypertrophic growth of the pancreas through this CCK-mediated negative feedback system (Green & Lyman, 1972). Although recent studies have suggested that this pathway is more complex than was originally proposed, the basic tenets of this concept of hormonal mediation are still valid.

The pancreatic enlargement caused by feeding rats with kidney bean diets has initially been ascribed to the presence of trypsin inhibitors in the seed (Greer *et al.*, 1985). However, a dose-dependent hypertrophy of the pancreas also occurs by an unknown mechanism when pure PHA is included in lactalbumin-based diets which contain no trypsin inhibitors (de Oliveira *et al.*, 1988). It is not known if the growth of the pancreas is mediated by CCK or other trophic factors but its enlargement shows some common features with the PHA-induced growth of the small intestine (Bardocz *et al.*, 1989*b*). Thus, the growth of both the small intestine and the pancreas is preceded by a highly significant accumulation of polyamines, particularly spermidine, in the tissue before any other changes are discernible. Most of this polyamine accretion is independent of *de novo* polyamine biosynthesis by ornithine decarboxylase (ODC). PHA is a growth factor for both the small intestine and the pancreas and its first effects are the stimulation of increased uptake of polyamines from circulating blood. However, it is not clear if stimulation of polyamine absorption into these tissues is a direct effect of the systemically absorbed PHA or if it is due to an indirect, perhaps hormonally mediated mechanism. A CCK-based mechanism appears unlikely as the pancreatic enlargement caused by trypsin inhibitors via CCK leads to increased insulin production (Bonnevie-Nielsen, 1981). In contrast, blood insulin levels fall when PHA causes hypertrophic changes in the pancreas (Pusztai, Grant & de Oliveira, 1986; Pusztai, 1986*a*).

The highly elevated urinary excretion of urea due to increased protein catabolism is one of the most striking effects of the dietary exposure to PHA or kidney bean proteins containing PHA on the systemic metabolism. It is possible that the high output of urea in the urine of rats fed on PHA-containing diets may be the result of PHA-induced loss of skeletal muscle. Both the wet and dry weights of the plantaris and gastrocnemius muscles decrease significantly with increased intake of dietary pure PHA (Table 6.6; de Oliveira *et al.*, 1988) and the reduction in muscle weight vs. body weight ratio increases progressively during PHA exposure. The reduction of the weight of both plantaris and gastrocnemius muscles by PHA-treatment is dependent on the intactness of the lectin function of PHA, since no such reduction occurs when PHA is replaced by a denatured preparation of the lectin in the diet (de Oliveira *et al.*, 1988). However, red muscles, such as the soleus, are not affected significantly within the 10 day experimental period.

Feeding rats with diets containing kidney bean proteins for 4 days leads to a reduction in the fractional rates of protein synthesis of the hind leg muscles *in vivo*, but the calculated fractional rates of protein degradation in the same muscles are unchanged. This leads to a net loss of plantaris and gastrocnemius muscle mass. As both the fractional rates of protein synthesis and degradation are increased to about the same extent in the soleus after treatment with PHA, the overall net effect of the lectin on this red muscle is slight. Although the reasons for these differences are unknown, the effects of PHA on the two types of muscles are clearly different (Palmer *et al.*, 1987).

After the first exposure to PHA, the rates of protein synthesis in both liver and hind leg muscles are initially unaffected in contrast to the immediate and dramatic increase in mucosal fractional protein synthesis rates after a first treatment with PHA (Palmer *et al.*, 1987). Accordingly, the chronic effects of PHA on muscles and the liver may be indirect and possibly hormonally mediated as are the overall catabolic effects of PHA on systemic metabolism and thymus atrophy. However, the results give only circumstantial evidence in support of hormonal mediation but there is some experimental evidence to indicate that PHA reduces the concentration of insulin in circulating blood and that the observed changes may be the result of this decreased level of insulin. The concentration of circulating immunoreactive insulin is depressed in rats fed on kidney bean diets for 10 days and it is maintained at this low level of 600–700 pg insulin/ml of serum continuously during the whole period of feeding (Pusztai, 1986*a,b*, 1989*a*; Pusztai *et al.*, 1986). This consistently depressed level of insulin in bean-fed rats is similar to that found in rats which are fed on protein-free diets (protein-free controls). In contrast, insulin levels are normal at about 2000 pg insulin/ml of serum in pair-fed control rats. Moreover, although the insulin concentration in rats pre-fed on kidney bean diets is low, further acute exposure to PHA by intragastric intubation induces an immediate and further drop in insulin concentration to values as low as 200 pg insulin/ml serum. It takes about 4–5 hours for the insulin concentration to recover to pre-exposure levels of about 600–700 pg/ml. According to preliminary results, the glucagon concentration in the serum changes in the opposite direction.

Pure PHA reproduces the insulin-lowering effects of kidney bean proteins and the inclusion of purified preparations of PHA isolectins of L_4 or E_3L in egg-albumin or lactalbumin-based diets results in a drop in the circulating insulin concentration similar to that obtained with kidney bean proteins containing similar amounts of the lectins (Fig. 6.9). These results are the first demonstration that the lectin component of kidney beans introduced into the small intestinal lumen reduces the concentration of an endocrine hormone, such as insulin, although the possibility that other components of the kidney bean seeds may also make a contribution to the overall reduction in insulin levels cannot be excluded. It is of interest that the L_4 isolectin appears to be more effective although both the L_4 and the mainly E subunit-containing E_3L PHA isolectins can reduce the insulin concentration.

The serum glucose levels in bean-fed rats are not increased despite the low insulin concentrations. Rather the opposite happens. These rats are

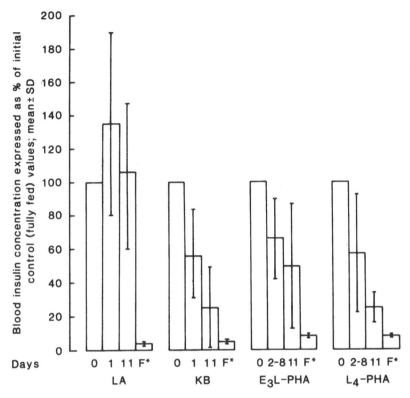

Fig. 6.9. Changes in circulating immunoreactive insulin concentrations in rats pair-fed on control (LA) diets or diets containing pure PHA-E$_3$L, PHA-L$_4$ or kidney bean proteins for 11 days, expressed as a percentage of blood insulin concentration of fully fed values ±SD before the start of the experiment. Rats were pre-fed on control 10% w/w lactalbumin (LA) diets (6 g/rat/day) for periods up to 10 days and blood-sampled from the tail vein three to four times before the start of the experiment. The animals were then selected into groups of six rats per treatment group and were given 6 g/rat/day control (LA) diet or a mixed lactalbumin (5% w/w) – kidney bean protein (5% w/w) diet (KB) or diets based on 9.3% lactalbumin containing 0.7% of the appropriate lectin (E$_3$L-PHA or E$_4$-PHA) for 11 days. The rats were sampled for blood 2 hours after the first exposure to the experimental diets (1) and then every second day for up to the 11th day of the experiment (2–8 and 11). Finally, all rats were fasted overnight and killed the next morning (F*). The results are given as means ± SD for the time points or are pooled values (2–8 days or 10–11 days) ± SD expressed as a percentage of the original fully fed values.

not diabetic despite the low blood insulin. It is possible that the high rates of depot lipid and muscle catabolism are both caused by the lack of insulin and these findings are consistent with the known effects of low insulin concentrations on metabolism. The involvement of insulin in mediating the effects of PHA is supported, since its induction of muscle catabolism is the result of a reduction in the fractional rates of protein synthesis rather than increased degradation.

The maintenance of normal blood glucose concentrations in rats which have been exposed to PHA for over three weeks also suggests that the production and/or secretion of other endocrine hormones may be affected in addition to changes in insulin concentrations. Indeed, changes in the concentration of glucagon do mirror those of insulin. The high rate of lipid catabolism also suggests that changes may occur in the level of glucocorticoids. It is likely that PHA modulates the overall hormone balance of the body, for the metabolic changes which occur during PHA exposure may reflect changes in most or all of the hormones produced by the endocrine system.

(d)   *Interference with local (gut) immune system; sIgA; IgE and systemic responses.* Macromolecules, including food proteins, can, and indeed, do cross the mature small intestinal epithelium and reach the systemic circulation (Pusztai, 1989c). In most instances this affects only a very small proportion of food proteins reaching the small intestine as most dietary proteins are quickly broken down by gut endopeptidases and the systemic absorption of such small amounts of intact proteins has no nutritional significance. However, this transepithelial transport is of great importance from the viewpoint of the defence of the body against harmful external agents. There are a number of possible and distinct routes for the uptake of macromolecules by the small intestine (Pusztai, 1989c). Of these, the transport of proteins through the membraneous epithelium of Peyer's patches is the most important for the immune-defence system and the gut-associated lymphoid tissues (GALT) are now generally regarded as the main compartment of the small intestine where sampling and presentation of antigens to the gut immune system occurs.

An observation in 1979 indicated that the orally presented lectin from kidney bean, PHA, is a very potent immunogen and produces an exclusive and powerful humoral, IgG-type, anti-lectin antibody response (Pusztai *et al.*, 1979b). It has since been shown that all other lectins of either plant or microbial origin, are also gut immunogens (de Aizpurua & Russel-Jones, 1988). Indeed, the humoral antibody response to lectins is similar regardless whether they have been presented parenterally or orally. In contrast, all non-lectin proteins or lectins which have been denatured by chemical or physical methods are only immunogenic when administered parenterally, and when given orally they elicit a minimal or nil antibody response.

Lectins differ from other food proteins in their high degree of resistance to proteolytic breakdown in the gut (see pp. 109–110). Appreciable amounts of lectins not only survive the gastrointestinal passage in an intact form, but also react with, and become endocytosed by, brush border cells. As a part of the intracellularly absorbed lectins is subsequently exocytosed from the epithelial cells through the basolateral membranes, lectins originally consumed with the diet may reach the systemic circulation. At low luminal concentrations most of the transepithelial transport of proteins is through M cells, but at high antigen concentrations absorption by the brush border also occurs through all epithelial cells (Walker, 1982). Clearly, even on the grounds of preferential survival, the amounts of systemically absorbed lectins are expected to be high as the luminal concentrations of intact lectins are high in comparison with those of non-lectin proteins. Indeed, an appreciable proportion of the PHA reaching the small intestine binds to, and is endocytosed by, all epithelial cells, including the most numerous enterocytes and goblet cells and not just by the M cells (King *et al.*, 1980*a,b*, 1982, 1986; Grant *et al.*, 1985; Pusztai 1989*a,b,c*). As lectins, and particularly PHA, are either partially resistant to lysosomal breakdown or may be routed through a non-lysosomal intracellular pathway after endocytosis (Pusztai, 1989*c*), an appreciable proportion of the intracellularly absorbed PHA may be exocytosed in fully functional form into the systemic circulation through the basolateral membranes. The amounts of systemically absorbed PHA may reach 5–10% of that initially given intragastrically (Pusztai, 1989*a,b*; Pusztai *et al.*, 1989*b*). However, with the less toxic tomato lectin, the absorbed proportion is less than 1% of that given initially (Kilpatrick *et al.*, 1985; Pusztai *et al.*, 1989*b*). The true significance of the extensive systemic absorption of PHA is clearly shown in that it occurs at a level three to four orders of magnitude higher than the amounts of non-lectin food protein antigens generally internalized by gut antigen-sampling through M cells (Baintner, 1986; Walker, 1982; Kleinman & Walker, 1984).

In addition to the systemic absorption of small, but significant amounts of PHA by transport through the columnar absorptive cells of the brush border epithelium, the interaction of the lectin with M cells may also be enhanced in comparison with other food proteins. As a result, PHA is transported into subepithelial tissues in the M cell regions of the small intestine, where the lectin is taken up by subepithelial macrophages and lymphocytes and carried into the systemic circulation (Fig. 6.10).

As shown by studies with [125]l-labelled PHA isolectins, the initial rate of lectin uptake by the gut cells is faster in rats which have been fed with kidney bean diets before the challenge with the radioactively labelled lectin. However, the extent of the systemic absorption of PHA becomes similar in all rats 3 h after the initial intragastric administration, including those which have been exposed to the lectin for the first time. Furthermore, the initial rate of uptake of the homotetramer PHA-L_4 exceeds that of the corresponding PHA-E_4 by a factor of nearly three, but the amounts of both types of PHA isolectins absorbed become the same after 3 h. The internalized PHA is not free in blood or other body fluids and cannot be removed from these by fetuin-Sepharose-4B, a selective absorbent that is generally used for the isolation of PHA by affinity chromatography (Pusztai & Palmer, 1977). Most of the lectin is bound to

(a)   (b)

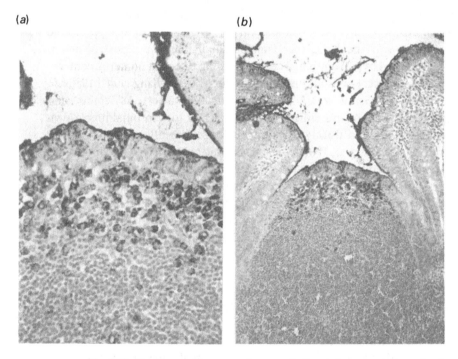

Fig. 6.10. The transport by macrophages and other antigen-presenting cells of pure PHA through small intestinal M cells into subepithelial tissues: (a) ×25 magnification; (b) ×10 magnification. Rats were fed on PHA-containing diets for 5 days. A 2 cm piece of the jejunum including Peyer's patch region was fixed in 10% neutral formalin and embedded in paraffin wax. After inhibition of the endogenous peroxidase, antigenic sites were unmasked with trypsinization and the sections were reacted with a solution of monospecific anti-PHA antibodies, followed successively by the link antiserum and peroxidase-anti peroxidase serum. The label was visualized with 3,3'-diaminobenzidine and sections were counterstained with haematoxylin (Ewen, Bardocz, Grant & Pusztai, unpublished observations).

serum glycoproteins and only small amounts are associated with blood cells, but, with time, the proportion of the PHA absorbed by blood cells increases. At first, this absorption is reversible and washing the cells with large volumes of buffers containing the specific haptenic glycoprotein, fetuin, removes the PHA. Later, however, an increasingly large proportion of the systemic PHA becomes irreversibly absorbed, suggesting that the lectin has been taken up by the blood cells by endocytosis (Table 6.7; Pusztai 1989*a*).

The original suggestion that a part of the lectin is reaching the circulation by transport through enterocytes and not only via M cells is supported by the observation that most of the systemically absorbed PHA is bound to circulating glycoproteins and not to cells. As most of the lectin is transported by macrophages and lymphocytes after uptake by the M cell route, the existence of glycoprotein-bound PHA argues strongly in favour of other types of transepithelial transport routes.

The identity of the blood cells which bind the systemically absorbed PHA has not yet been established. Even when fed on kidney bean proteins instead of on purified PHA, all animal species tested develop a powerful and selective humoral antibody response of the IgG type to the dietary lectin (Pusztai *et al.*, 1979*b*, 1981*b*; Grant *et al.*, 1985; King, Begbie & Cadenhead, 1983; Begbie & King, 1985; Williams *et al.*, 1984), hence the blood cells must have included fully immunocompetent lymphocytes (Fig. 6.11). In fact, the time-course of antibody development shows the usual features of immunization, and a measurable primary response has been found after about 10 days from the initial exposure to PHA. Even after relatively long periods of feeding on lectin-free diets, re-introduction of PHA into the diets of rats after the initial exposure and/or further periods of feeding with kidney bean diets results in considerable increases in the anti-lectin antibody titre. Accordingly, re-exposure to the sensitizing antigen, PHA, produces considerable secondary, booster effects (Grant *et al.*, 1985). Even when the rats or pigs are fed with unpurified kidney bean meal, the observation that the antibody produced is monospecific for PHA highlights the great specificity and selectivity of the transepithelial transport of the lectin in contrast to the exclusion of all other bean proteins or antigens. Additionally, the immunosuppressant effects of PHA to other antigens presented simultaneously may also contribute to the exclusiveness of the response of the humoral immune system to PHA (see p. 95).

The occurrence of powerful booster effects when PHA is re-introduced into rat diets, indicates that the potential neutralizing effects of the local antibody system, which is based on putative anti-lectin sIgA,

Table 6.7. *Survival and uptake into systemic circulation of ^{125}I-labelled lectins given orally to rats.*

| | PHA | | | | Tomato lectin |
| | E3L | | L4 | | |
	1 h (n = 11)	3 h (n = 11)	1 h (n = 6)	3 h (n = 4)	3 h (n = 2)
Small intestine					
contents[a]	16.44 ± 5.50	16.00 ± 1.45	7.10 ± 0.73	3.30 ± 0.97	15.82
tissue bound[a]	14.15 ± 0.43	8.70 ± 2.54	8.07 ± 1.33	4.84 ± 0.83	8.00
Caecum + colon[a]	0.34 ± 0.18	0.64 ± 0.11	0.25 ± 0.14	0.54 ± 0.09	15.00
Blood cell bound					
(from 1 ml)					
Elutable[a]	0.06 ± 0.01	0.14 ± 0.04	0.16 ± 0.03	0.26 ± 0.04	<0.01
Non-elutable	0.04 ± 0.01	0.06 ± 0.01	0.07 ± 0.02	0.11 ± 0.02	<0.01
Plasma (1 ml)[a]	0.19 ± 0.04	0.57 ± 0.07	0.60 ± 0.06	0.43 ± 0.15	0.01
Organs					
Liver[a]	0.65 ± 0.11	0.61 ± 0.01	n.d.	n.d.	0.71
Kidney[a]	0.11 ± 0.02	0.16 ± 0.02	n.d.	n.d.	0.10
Pancreas[a]	0.07 ± 0.02	0.04 ± 0.01	n.d.	n.d.	0.01
Spleen[a]	0.04 ± 0.01	0.01 ± 0.01	n.d.	n.d.	0.02

Note: [a] Trichloroacetic acid-precipitable activity.

Results are expressed as a mean percentage (±SD) of the dose reaching the small intestine. E3L and L4 lectins were prepared (Manen & Pusztai, 1982), labelled with ^{125}I (Kilpatrick et al., 1985) and purified on the appropriate antibody-Sepharose 4B column (Manen & Pusztai, 1982). Fasted rats were given 380 µmol of sodium iodide by intravenous injection. After 30 min, 1 µCi of labelled lectin, dissolved in 1 ml saline solution containing 50 mg of kidney bean extract was given intragastrically. Counts of radioactivity (protein-bound, insoluble in 5% trichloroacetic acid) in blood and internal organs were measured. n.d., not determined.

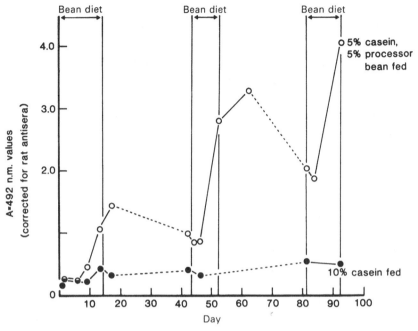

Fig. 6.11. The time-course of development of systemic anti-PHA, IgG-type, antibodies in rats fed alternately for periods of 5 days on kidney bean proteins or control (LA) diets. Rats were blood-sampled throughout the experiment and the concentration of anti-PHA antibodies was measured by an ELISA assay. The reaction was visualized with 3,3′-diaminobenzidine after the anti-PHA antibody from the rat serum, fixed to the PHA-coating of the wells of the microtitre plate, was reacted with peroxidase-labelled anti-rat IgG antibody (from Grant *et al.*, 1985; by permission of *The Journal of the Science of Food and Agriculture*).

must be relatively ineffective since it cannot prevent the systemic absorption of at least some of the PHA introduced into the small intestine by repeated feedings. It is, however, not known whether this relative ineffectiveness of the local antibody to neutralize luminal PHA is due to the relatively poor memory of the sIgA antibody system or whether PHA generally interferes with the production of IgA and sIgA antibodies. As feeding rats with PHA diets is usually accompanied by a dramatic proliferation of bacteria in the small intestine (Jayne-Williams & Hewitt, 1972; Rattray, Palmer & Pusztai, 1973; Wilson *et al.*, 1980; Banwell *et al.*, 1983, 1985), it is possible that PHA may generally suppress the proper functioning of the sIgA-based local immune system.

In contrast to the largely circumstantial evidence of disturbances caused by PHA in the functioning of the sIgA-based gut immune system, the experimental evidence is more substantial that PHA affects mucosal mast cells leading to anti-lectin IgE production in the small intestine. The anaphylactic effects of PHA on gut mucosal mast cells (p. 102) and the consequent exudation and loss of serum proteins into the gut lumen may be the result of the direct de-granulation of mast cells by PHA traversing the intestinal wall (Pusztai & Greer, 1984; Greer & Pusztai, 1985). Longer dietary exposures to PHA elevate this anaphylactic response appreciably. After a period of lectin-free diets, the existence of a memory effect in response to the re-introduced lectin or to challenge-doses of PHA suggests that the initial non-immune anaphylaxis may be amplified by the production of anti-PHA IgE-type antibodies on extended oral exposure to the lectin. The loss of serum proteins into the intestine increases mainly as a result of an elevated mucosal, vascular permeability. This local IgE-based allergenic immune reaction contributes to the overall poor nutritional state of the body and aggravates the nutritional toxicity of PHA.

In addition to local hypersensitivity reactions to dietary PHA, both mice and rats develop systemic, immediate-type (Type-1) hypersensitivity due to the production of systemic anti-lectin IgE (Pusztai *et al.*, 1983*b*). Mice are sensitized to dietary PHA and this can be demonstrated by the booster effects in IgE production after the intraperitoneal injection of PHA with $Al(OH)_3$ as adjuvant. Similar, though less extensive booster effects have also been observed after the oral re-administration of PHA 10 days after the initial exposure. Although the extent and the nature of systemic allergic responses and their contribution to the antinutritional effects of PHA are not known, as some lectins, such as PHA or jacalin, suppress IgE response to other simultaneously applied antigens (Restum-Miguel & Prouvost-Danon, 1985), consumption or therapeutic application of lectins may be used to alleviate allergic responses in sensitive people.

(e)   *Modulation of the microbial ecology of oral cavity and small intestine.* The effects of bacteria on the functioning of the digestive tract are complex even when infections by pathogenic species in the oral cavity and the upper parts of the gastrointestinal tract are not considered. Also, the interactions of food lectins with bacteria resident in the large intestine are unknown. Therefore, the dietary modulation of the microbial flora in parts of the alimentary canal distal to the ileum is outside the scope of this book.

With the exception of starchy polysaccharides, the degradation of components of food is minimal in the oral cavity, but through chewing, mastication and thorough mixing with saliva, it is prepared here for the breakdown which occurs later in the stomach and the small intestine. The demonstration, that hot or cold aqueous extracts from a number of commonly ingested fruits, vegetables and seeds react with human saliva and agglutinate the cells of *Streptococcus mutans* strains (Gibbons & Dankers, 1981), is in keeping with the generally accepted notion that components of food pre-determine the microbial ecology and health of the oral cavity. The extracts prepared from these foods can also inhibit the attachment of the cells of *Streptococcus mutans* H12 or *Streptococcus sanguis* c1 strains to saliva-coated hydroxyapatite beads, which are used as model teeth in dental studies. As these interactions are abolished by sufficiently high concentrations of specific sugars, the observations suggest the involvement of food lectins in host–parasite type interactions in the mouth (Gibbons & Dankers, 1981). Indeed, saliva and salivary mucins selectively inhibit the binding of food lectins to teeth and buccal epithelial cells. Accordingly, it is possible that one of the main functions of the salivary secretions of the oral cavity is to reduce the extent of this binding of food lectins to teeth and, through this inhibition, to reduce the extent of the harmful agglutination and adherence of bacteria to their surface (Gibbons & Dankers, 1982). The generally high reactivity of food lectins with the oral mucosa both *in vivo* and *in vitro* (Rittman, Mackenzie & Rittman, 1982; Gibbons & Dankers, 1983; Hietanen & Salo, 1984) suggests that the attachment of increased numbers of streptococci to epithelial cells is a potentially important part of the overall effects of food on the bacterial ecology of the oral cavity.

Arguably, the effects of bacteria present in the small intestine are of great importance for its proper functioning. The presence of bacteria in the small intestine may be harmful in a number of instances in contrast to the large numbers of resident bacteria in the large intestine, where they are adapted to function under normal conditions without any deleterious effects for the animal. Nevertheless, some animals, such as rabbits or rats, may acquire vitamins, essential micronutrients or even enzymes through coprophagy, and animals may also derive some benefit in the presence of relatively innocuous bacteria in the small intestine if this leads to the exclusion of potentially pathogenic species. As both bacterial and small intestinal digestive processes compete for the same food eaten, bacterial overgrowth of even non-pathogenic strains may reduce the amounts of nutrients for the animals.

Bacterial counts in the small intestine of healthy animals are usually low, particularly in laboratory animals, and the extent of competition of bacteria with the host's digestive system is negligible. However, it is becoming increasingly clear that the most important interaction between food components and bacteria is not simple competition for food. Food and some of its components, particularly the lectins, may directly interact with the bacterial flora or, alternatively, they may indirectly affect bacterial proliferation in the small intestine through interference with the binding of selected species to epithelial tissues. Whichever of these two mechanisms operates, the end result is the potential inducement of selective proliferation of some species of bacteria in the digestive tract, including the small intestine.

Bacterial binding (Sharon, 1987) or the attachment of parasitic protozoa (Lev *et al.*, 1986) to surface epithelial cells is through their specific lectin-adhesins. The extent of this binding may be modified by components of food or gut secretions which thus may induce a selective proliferation of the appropriate organisms. As these interactions may be competitive, additive or synergistic (Fig. 6.12), the bacterial ecology, the barrier function and the efficiency of food assimilation are dependent on the composition of the diet. The extent of the interference with the digestion/absorption of food components is, in turn, dependent on the severity of the bacterial overgrowth. However, in addition to the increased competition for food between the small intestine and the proliferating bacteria, the presence of large numbers of microorganisms adhering to the mucosa may lead to destructive changes in normal small intestinal morphology. These changes may result in further deterioration of the efficiency of digestion and absorption of food.

The damaging contribution of bacteria to the already poor nutritional utilization of kidney bean diets is shown by the dramatic differences in the effects of PHA-containing diets on conventional vs germ-free animals (Table 6.8). Thus, raw kidney bean diets are far less toxic for gnotobiotic than for conventional animal species as different as rats (Rattray, Palmer & Pusztai, 1974) and quails (Jayne-Williams & Hewitt, 1972). Furthermore, concurrent with increased toxicity, there is a dramatic overgrowth of *Escherichia coli* in the small intestine of conventional rats fed on PHA-containing diets (Wilson *et al.*, 1980). When the amounts of PHA in the diets of rats are reduced substantially by feeding kidney beans of a low-lectin variety, Pinto III, this overgrowth does not occur and diets based on Pinto III bean are also far less toxic for rats than those containing high-lectin kidney bean cultivars (Wilson *et al.*, 1980).

Similar studies with other animal species have fully confirmed the existence of this causative relationship between the presence of PHA in the diet, *E. coli* overgrowth and toxicity (Jayne-Williams & Burgess, 1974; Banwell *et al.*, 1984). Although the mechanism of the selective overgrowth and how this affects nutritional efficiency is not clear, one possible mechanism is that the lectin-induced virulence of coliforms in the small intestine of kidney bean-fed rats is the result of the elimination of competing species. Lectins may also facilitate the agglutination of bacterial cells to each other and/or to mucosal surfaces and, thus, give selective advantage for the growth of individual species (Wilson *et al.*, 1980). As proliferating cells of *E. coli* may bind to mannose-containing receptors through their mannose-specific fimbrial adhesins, attachment of the coliforms may be partly or even exclusively to the polymannose side-chains of PHA which are already bound to the microvilli (King *et al.*, 1983; Banwell *et al.*, 1984; Boldt & Banwell, 1985). However, the evidence is against most of these proposed mechanisms of PHA-induced proliferation of the bacteria since no specific association between PHA and the bacteria has been found by bacterial agglutination assays or glycocalix stabilization. The extent of bacterial adherence to washed small intestinal mucosa was not increased in the presence of PHA in three different binding assays and the conclusion appears to be inescapable that PHA probably does not serve as a ligand to mediate bacterial adherence to mucosal surfaces (Ceri *et al.*, 1988).

Despite the lack of a clear understanding of the mechanism of the promotion of bacterial overgrowth by PHA, the occurrence of a reversible and PHA-dependent bacterial overgrowth has been demonstrated repeatedly. The time-course and the population dynamics of microbial colonization of the small intestine by major aerobe and facultative anaerobe groups have been studied recently in PHA-fed rats (Banwell *et al.*, 1988). The predominant bacterial isolates contain *E. coli*, a *Streptococcal* sp. and *Lactobacillus* and bacterial populations increase within 24 hours of the first exposure to PHA and this increase is sustainable on further exposure to the lectin. On the other hand, when PHA is taken out of the diet, coliform counts revert to control levels within 24–48 hours and gram-positive rods and cocci become dominant in the flora. In addition, the *E. coli* isolated from the PHA-induced overgrowth in the small intestine express no predominant serotype or fimbriae and no isolates contain either heat-stable or heat-labile toxins. Thus, it has been suggested that the effects of PHA may be indirect and mediated by the destructive changes in the morphology and composition of mucosal surfaces caused by the surface-bound dietary lectin (Ceri *et al.*, 1988; Banwell *et al.*, 1988).

Bacterial proliferation occurs very rapidly after the first exposure to PHA. Within 24 to 48 h, the initial coliform counts of about 10^3–10^4 organisms/g of wet small intestinal tissue of the Rowett's colony of Hooded-Lister rats increase to about 10^{10}–10^{11} (Wilson *et al.*, 1980).

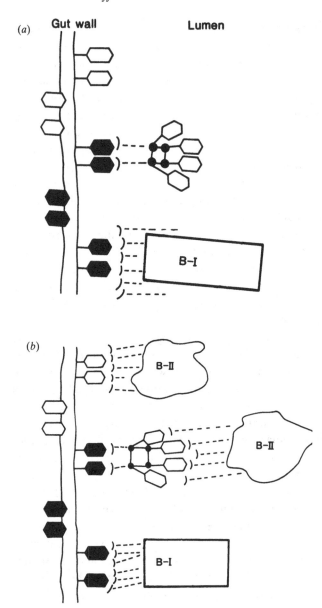

Fig. 6.12(a,b)

These bacterial counts are almost as high as those in ruminants. Although the binding of PHA to the brush border epithelium is fast, disruption of its structure occurs somewhat more slowly and may need several days to reach its peak. Similarly, while the binding of PHA is practically instantaneous, bacterial counts increase gradually over several days. There is

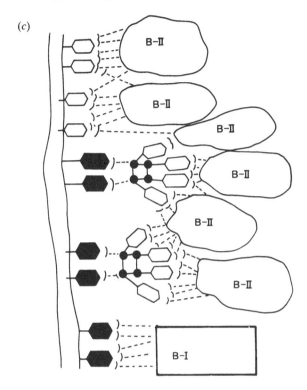

Fig. 6.12. Schematic diagrams of possible and different types of interactions between food lectin, two different bacterial adhesins (B-I and B-II) and carbohydrate receptors of small intestinal epithelium: (*a*), competitive; (*b*), additive and (*c*), synergistic interactions. The scheme also implies that dietary carbohydrates appropriate for the specificity of the lectins or adhesins may be potential inhibitors of the binding of food lectins or are blockers of bacterial adhesion.

no evidence to suggest that bacterial proliferation is the result of the accumulation of sloughed off cells and other cellular debris in the lumen of the small intestine. In contrast, while the binding of *E. coli* to mucosal surfaces is slight in control rats, there is an avid bacterial binding to the epithelium both *in vivo* and *in vitro*, as if the access of the bacteria to epithelial surfaces was increased in the presence of PHA (Ceri *et al.*, 1988). Considering that the secretion of mucinous glycoproteins increases after exposure to PHA and to some other kidney bean proteins, it seems likely that the access of the bacteria to epithelial surfaces may be reduced rather than increased by entrapment of the bacteria in the mucin coat and their continuous removal by peristalsis.

The speed of bacterial proliferation suggests that PHA may directly promote the synthesis of new receptors on epithelial surfaces which are

Table 6.8. *Nutritional performance of* (a) *conventional or* (b) *germ-free rats fed on raw kidney bean diets.*

Diet	Food intake (g/day)	Weight change (g/day)	Digestibility (true; %)	NPU (%)
(a)				
Casein	8.8	+3.3	98	90
Bean (½) + Casein (½)	4.1	−0.9	61	20
Bean	2.8	−5.3	All rats died after 3 days	
(b)				
Casein	6.9	+2.3	98	85
Bean (½) + Casein (½)	4.9	+1.5	85	66
Bean	4.3	+0.2	73	29

Note: Rats (groups of four) were fed *ad lib* on the appropriate diets for 10 days. Daily food intakes and weight changes per four rats are given. True digestibility and the net utilization of dietary protein (NPU) over the 10-day period were measured from the nitrogen content of faeces and body.

capable of binding of bacteria. This is more in keeping with the known and almost immediate effects of the endocytosed PHA on increased protein, RNA and DNA synthesis in epithelial cells (see p. 110). There are also indications that the extent of endocytotic uptake of PHA by epithelial cells of the proximal small intestine is substantially reduced in germ-free rats although the reasons for this are not understood. Rather curiously, PHA binds to the mucosa and causes destructive changes in the morphology of the epithelium, which are very similar to that occurring in conventional rats (Figs 6.13 & 6.14). However, the immunofluorescence patterns of small intestinal sections of germ-free rats fed on kidney bean diets show a conspicuous lack of penetration of PHA into the epithelial cells (Pusztai *et al.*, 1990*b*), in accord with results obtained by nutritional testing, which have shown that PHA is not lethally toxic for gnotobiotic animals (Rattray *et al.*, 1974), although kidney bean diets are not utilized with great efficiency even by these rats. It can be assumed as a working hypothesis that there is a correlation between the reduced toxicity of PHA for germ-free rats and the decreased rates of endocytosis of the toxic lectin by the cells of the small intestine. It is possible that the epithelial cell membrane of the small intestine of conventional rats fed on PHA-containing diets may become progressively more destabilized by the combined effects of PHA and the adhesions of *E. coli* than by the attachment of the dietary lectin alone. As the two lectins bind to different

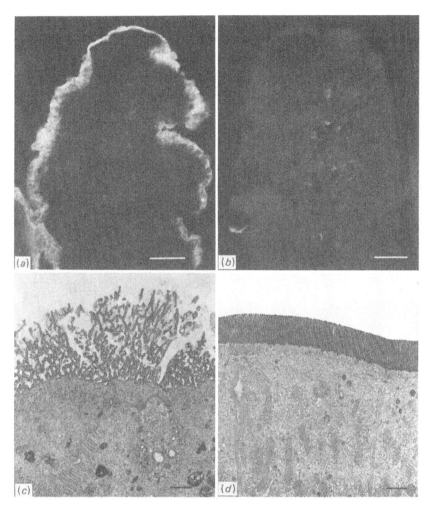

Fig. 6.13. Immunofluorescence and electron micrographs of a part of a section through the jejunum of germ-free rats fed on diets containing kidney bean or casein controls. (*a*) Immunofluorescence obtained with a section of the jejunum incubated with rabbit anti-lectin immunoglobulins and FITC-conjugated anti-rabbit IgG. Strong labelling is shown in the brush border regions. However, by comparison with similar immunofluorescence staining of the jejunum of conventional rats (King, Pusztai & Clarke, 1980*a*,*b*), the level of intracellular labelling is much reduced. Scale bar: 25 μm. (*b*) Control; a section from the same specimen as in (*a*), but incubated with non-specific rabbit immunoglobulins and FITC-conjugated anti-rabbit IgG shows no staining of the brush border. Scale bar: 25 μm. (*c*) Electron micrograph of a part of a section through the apical region of a jejunal enterocyte from a germ-free rat fed on kidney bean diet. Severe disruption of microvilli and some intracellular changes are shown. Scale bar: 1 μm. (*d*) A similar section as in (*c*) obtained from the jejunum of a rat fed on control (casein) diet showing the normal appearance of microvilli. Scale bar: 1 μm.

receptors, the combined (Fig. 6.12) strain on the membrane, which may be synergistic, may lead to enhanced endocytic uptake of PHA or of both lectins. In contrast, with the reduction in the intracellular absorption of the lectin, the very damaging and wasteful effects of compulsory growth of the small intestine will be largely eliminated in gnotobiotic rats. Also, the reduced extent of systemic absorption will abolish or reduce the damaging and hormonally mediated catabolic effects of PHA on general systemic metabolism.

Bacterial proliferation in the small intestine affects the immune function of the gut and although cell-mediated and other immune reactions which follow PHA-induced bacterial proliferation may aggravate the injurious direct effects of PHA, these are indirect effects and not discussed further.

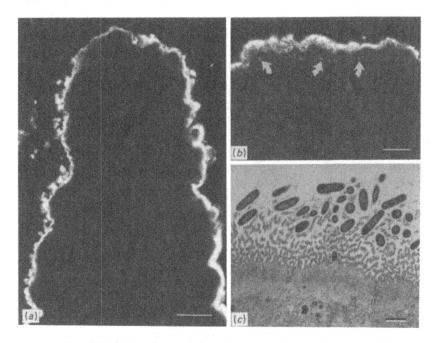

Fig. 6.14. Immunofluorescence and electron micrograph of parts of jejunal sections of conventional rats fed on diets containing kidney beans. (*a*) Immunofluorescence obtained with a section of the jejunum incubated with rabbit anti-lectin immunoglobulins and FITC-conjugated anti-rabbit IgG. Strong labelling is shown in the brush border regions. Scale bar: 25 μm. (*b*) Similar specimen as (*a*) but at slightly higher magnification to reveal the presence of dietary kidney bean lectin within intracellular vesicles. Scale bar: 50 μm. (*c*) Electron micrograph showing overgrowth of coliform bacteria associated with the damaged microvilli caused by kidney bean lectins. Scale bar: 1 μm.

Effects of other lectins on cells of the small intestine

Soyabean agglutinin. For a major part of this century it has been known that the nutritional value of raw soyabeans is enhanced by proper heat treatment. There have been many reasons suggested as to the possible causes for the poor growth and other nutritional problems encountered when monogastric animals are fed on diets based mainly or exclusively on raw soyabeans (Jaffè, 1980; Liener, 1986). At various times it has been proposed that seed constituents such as the two different trypsin inhibitors, the lectin, saponins or polyphenols and tannins act either singly or in combination as antinutritive agents. There is no firm experimental evidence to support the suggestions that a single component can satisfactorily explain all the effects of soyabean diets on growth retardation and systemic metabolism. In fact, the evidence indicates quite the opposite since purified trypsin inhibitors tested at the concentration found in the soyabean meal can only account for about half of the growth inhibition. Similarly, the soyabean lectin cannot itself account for all the antinutritive effects of soyabean. Indeed, the removal of the glycoprotein lectin (and other glycoprotein seed components) by affinity chromatography on concanavalin A-Sepharose-4B does not abolish the growth retardation effect of raw soyabean-feeding (Turner & Liener, 1975). Nevertheless, the soya lectin, or other potentially deleterious seed components have definite antinutritive effects even when tested singly. For example, partially purified soyabean agglutinin preparations have damaging effects on intestinal brush border enterocytes, similar to those caused by PHA (Jindal, Soni & Singh, 1984). However, although such effects are in keeping with lectin injury generally, the soya lectin preparations used by Jindal *et al.* (1984) have been rather impure, with a purification factor of less than elevenfold. Their results are therefore unacceptable as firm evidence for the involvement of soyabean agglutinin as the cause of the antinutritive effects of soyabean. It is likely that the growth retardation effects obtained by Jindal *et al.* (1984) were due to the presence of several factors in the preparation. Jaffè (1980) suggested from the experimental evidence available at the time that growth inhibition by soyabeans is likely to be the result of complex interactions between a number of antinutritive factors, such as the lectin, trypsin inhibitors and results obtained in the last ten years have confirmed this view.

The lectin and saponin preparations obtained from soyabean have a mild destabilizing effect *in vitro* on the everted rabbit jejunum when tested separately (Alvarez & Torres-Pinedo, 1982). When applied simultaneously, there is a 100-fold amplification in the permeability of the jejunum in comparison with that found with either of the antinutrients used alone. Thus, the destabilizing effects of the two components are not additive but synergistic (Torres-Pinedo, 1983).

There is also evidence from morphological studies with small bowel biopsies obtained from infants and small children suffering from soyabean-related allergies, that the villus atrophy and other structural lesions found in the small intestinal epithelium resemble lectin-induced injuries and toxicity (Perkkio, Savilahti & Knitunen, 1981; Poley & Klein, 1983). These lesions are similar to those seen in the small intestine of coeliac people with gluten-induced enteropathy. Soyabean agglutinin also binds to fixed sections of the small intestinal epithelium (Ovtscharoff & Ichev, 1984). However, as this binding is to crypt cells, it is difficult to reconcile it with the *in vivo* villus injuries and atrophy found in the small bowel biopsies from children with soyabean intolerance.

Although the soyabean agglutinin is not solely responsible for the poor nutritional value of soyabeans, it definitely has antinutritive effects (Grant *et al.*, 1986, 1987*a*,*b*,*c*). Pure soyabean lectin inhibits the growth of rats and induces very considerable changes in the morphology and the functioning of both small intestine and pancreas. In a very similar way to that found with PHA-containing diets, pure preparations of soyabean agglutinin induce hyperplastic growth of the small intestine in a dose- and time-dependent way (Grant *et al.*, 1987*c*; Pusztai *et al.*, 1990*a*). A comparison of the effects of various lectins on the mean weight gain of rats (Fig. 6.15) or the mean dry weight of their jejunum (Fig. 6.16) shows that the lectin from soyabean is slightly less effective as a growth factor for the small intestine than PHA. This is borne out by further comparisons of the mean villus/crypt length (Fig. 6.17) and of the mean villus/crypt cell count (Fig. 6.18). In a further analogy to PHA-induced growth, the proliferation of crypt cells occurring after the gut lumen is exposed to pure soya lectin, is dependent on a coincident and highly significant increase in tissue polyamine levels. Although this is partly ameliorated by treatment with α-difluoromethylornithine (DFMO, an irreversible inhibitor of ornithine decarboxylase, ODC, the rate-limiting enzyme of polyamine biosynthesis) carried out concurrently with the lectin treatment, crypt cell proliferation is not abolished. The reason for this may be that polyamines needed for growth are taken up from circulation and not synthesized in the tissue to any great extent as also found with the PHA-induced hyperplastic growth of the small intestine. As the basolateral uptake of polyamines is insensitive to DFMO, the presence of this ODC inhibitor has only a slight effect on cell proliferation (Bardocz *et al.*, 1990*a*,*b*).

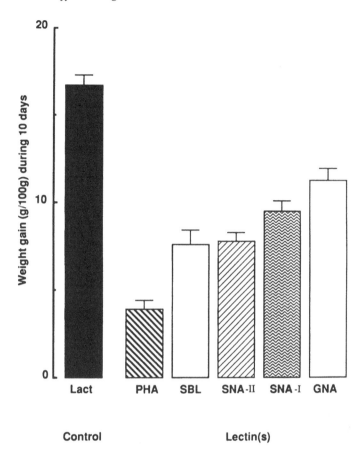

Control   **Lectin(s)**

Fig. 6.15. Mean weight gain of rats fed for 10 days on diets containing 9.3% (w/w) lactalbumin and 0.7% of various individual lectins (5.7 g of diet/rat/day). Control rats were given the same amount of a control diet containing 10% (w/w) lactalbumin. The results, mean values ± SD, are compared with those obtained with control rats fed on lactalbumin-diets (Lact) by statistical analysis, Student t test. The following lectins were included: PHA (*Phaseolus vulgaris*); SBL (soyabean, *Glycine max*); SNA-I and SNA-II, from elderberry (*Sambucus nigra*) bark and GNA, snowdrop (*Galanthus nivalis*) bulb lectin.

Dietary exposure to soyabean agglutinin also causes a highly significant enlargement of the pancreas (Figs 6.19 & 6.20), which is not very surprising as the cellular origin of the pancreas is similar to that of the small intestine. The growth is initially hypertrophic and dependent on the accumulation of polyamines in the tissue (Grant *et al.*, 1989; Pusztai *et al.*, 1990*a*). The total polyamine content of the pancreas is significantly increased within 6 hours of the initial dietary exposure to soyabean and then it further increases progressively throughout the 16 day experimental period. In the first 6 hours there are no significant changes in tissue weights and/or DNA, RNA and protein contents. Accordingly, the accumulation of polyamines (mainly spermidine) in the tissue precedes the growth. It appears that most of the polyamine accretion is due to uptake from blood circulation, as the activity of ODC in the pancreatic

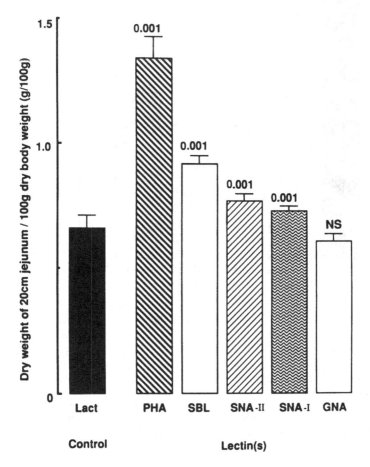

Fig. 6.16. Mean dry weight of a 20 cm piece of jejunum of rats fed on diets containing 0.7% of various individual lectins and 9.3% lactalbumin (5.7 g of diet/rat/day) or the same amounts of control diet (10% lactalbumin). All measurements were made on the same rats (groups of four) as in Fig. 6.15. The results, mean values ± SD, are compared with those obtained with control rats fed on lactalbumin-diets (Lact) by statistical analysis, Student t test. Significant differences (with p values) are denoted above the bars of standard deviation (SD). NS is not statistically significant. The following lectins were included: PHA (*Phaseolus vulgaris*); SBL (soyabean, *Glycine max*); SNA-I and SNA-II, from elderberry (*Sambucus nigra*) bark and GNA, snowdrop (*Galanthus nivalis*) bulb lectin.

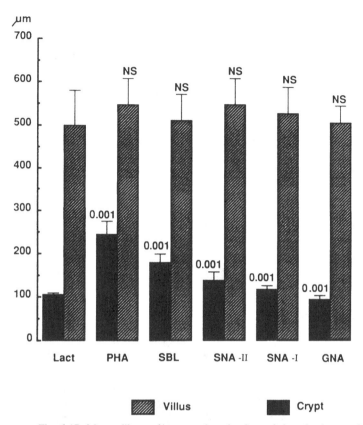

Fig. 6.17. Mean villus and/or crypt length of rats fed on lectin-containing or control diets. Ten villi or crypts selected at random from each rat were measured. All measurements were made on the same rats (groups of four) as in Fig. 6.15. The results, mean values ± SD, are compared with those obtained with control rats fed on lactalbumin-diets (Lact) by statistical analysis, Student t test. Significant differences (with p values) are denoted above the bars of standard deviation (SD). NS is not statistically significant. The following lectins were included: PHA (*Phaseolus vulgaris*); SBL (soyabean, *Glycine max*); SNA-I and SNA-II, from elderberry (*Sambucus nigra*) bark and GNA, snowdrop (*Galanthus nivalis*) bulb lectin.

tissue is low and does not increase during growth. Moreover, DFMO does not inhibit the soya agglutinin-induced pancreatic enlargement (Grant *et al.*, 1989).

The anti-nutritive effects of soya lectin are now well established. Its binding to the small intestinal epithelium induces a number of changes in intestinal function and morphology (Fig. 6.21) some of which reduce the food conversion efficiency in soya-fed rats. Because of possible synergisms with other antinutrients in the seed meals, the growth retardation and other damaging effects of dietary exposure to soyabean agglutinin are likely to be variable and dependent on the conditions of testing. However, most soya preparations used in feeding studies contain appreciable amounts of functional lectin (Prince *et al.*, 1988). Moreover, brush border membranes have many binding sites for the soya agglutinin (Hendriks *et al.*, 1987). Thus, potentially damaging interactions between the small intestinal epithelium and the dietary soya lectin may be of a regular daily occurrence in both humans and animals.

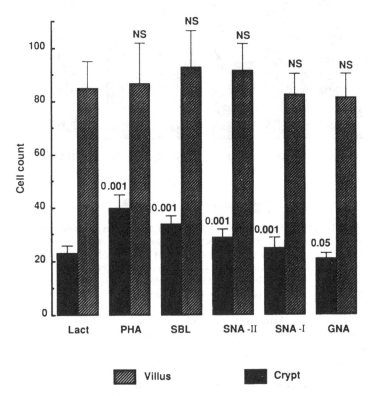

Fig. 6.18. Mean villus/crypt cell counts; 10 villi/crypts selected at random from each rat were counted. All measurements were made on the same rats (groups of four) as in Fig. 6.15. The results, mean values ± SD, are compared with those obtained with control rats fed on lactalbumin-diets (Lact) by statistical analysis, Student t test. Significant differences (with p values) are denoted above the bars of standard deviation (SD). NS is not statistically significant. The following lectins were included: PHA (*Phaseolus vulgaris*); SBL (soyabean, *Glycine max*); SNA-I and SNA-II, from elderberry (*Sambucus nigra*) bark and GNA, snowdrop (*Galanthus nivalis*) bulb lectin.

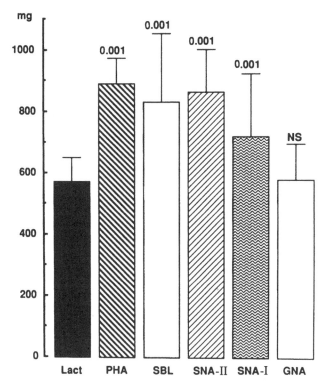

Fig. 6.19. Mean pancreatic wet weight of rats (groups of four) fed (5.7 g/rat/day) on diets containing 0.7% (w/w) of various individual lectins and 9.3% (w/w) lactalbumin or control diet (10% lactalbumin). The results, mean values ± SD, are compared with those obtained with control rats fed on lactalbumin-diets (Lact) by statistical analysis, Student t test. Significant differences (with p values) are denoted above the bars of standard deviation (SD). NS is not statistically significant. The following lectins were included: PHA (*Phaseolus vulgaris*); SBL (soyabean, *Glycine max*); SNA-I and SNA-II, from elderberry (*Sambucus nigra*) bark and GNA, snowdrop (*Galanthus nivalis*) bulb lectin.

Concanavalin A. Jack bean (*Canavalia ensiformis*) is not used extensively in animal nutrition and, although the involvement of concanavalin A in antinutritional effects has not been thoroughly tested, it is known that the meal will support the growth of rats only after proper heat treatment. However, it is questionable whether the pathological damage found in these rats is all due to concanavalin A in the seed meal fed to the animals (Liener, 1986).

In contrast, the effects of concanavalin A on the small intestine *in vitro* have been studied extensively. Thus, concanavalin A binds to brush border membranes (Etzler & Branstrator, 1974; Ichev & Ovtscharoff, 1981) and exposure of the small intestinal lumen to concanavalin A leads to injury, some of which is rather similar to that seen with PHA (Lorenzsonn & Olsen, 1982; Ichev & Ovtscharoff, 1981). The lesions induced by concanavalin A introduced into small intestinal loops include increased shedding of brush border membranes, reduction of the surface area of the small intestine, accelerated cell loss and slight villus atrophy. Most of these damaging changes can be prevented by the simultaneous administration of high enough concentrations of the appropriate haptenic sugar, D-glucose (Lorenzsonn & Olsen, 1982). Furthermore, there are indications that the transport of amino acids is inhibited by concanavalin A in the lumen. The lectin may also interfere with the synthesis of small intestinal protein and the proper intracellular distribution of newly synthesized proteins (Ichev & Chouchkov, 1983). On the basis of the effects of concanavalin A and wheatgerm agglutinin, Lorenzsonn and Olsen (1982) have suggested that dietary lectins may be, in part, respon-

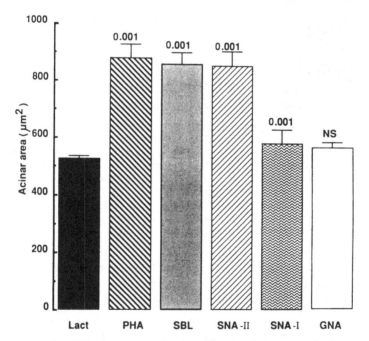

Fig. 6.20. Mean pancreatic acinar area of rats fed on various lectin or control diets as in Fig. 6.19. The results, mean values ± SD, are compared with those obtained with control rats fed on lactalbumin-diets (Lact) by statistical analysis, Student t test. Significant differences (with p values) are denoted above the bars of standard deviation (SD). NS is not statistically significant. The following lectins were included: PHA (*Phaseolus vulgaris*); SBL (soyabean, *Glycine max*); SNA-I and SNA-II, from elderberry (*Sambucus nigra*) bark and GNA, snowdrop (*Galanthus nivalis*) bulb lectin.

sible for the high natural turnover of brush border membranes under normal conditions.

Concanavalin A has exceptionally high resistance to small intestinal proteolysis during its passage through the alimentary canal as has PHA; over 90% of dietary concanavalin A can be recovered intact from the faeces (Nakata & Kimura, 1985). The binding of the lectin to surface epithelial cells of the small intestine also causes disturbances in the formation and re-formation of brush border membranes (Nakata & Kimura, 1986).

The hyper-regenerative adaptation of the rat intestinal mucosa to the cytotoxic effects of the lectins found previously by Lorenzsonn & Olsen (1982), has been fully confirmed in detailed studies (Lorenz-Meyer *et al.*, 1985) of the effects of the introduction of concanavalin A, or wheatgerm agglutinin, into perfused, self-emptying jejunal loops. Concanavalin A (and to some extent, wheatgerm agglutinin) reduces villus height and causes a deepening of the crypts, and there is an excessive increase in cell exfoliation at the villus tips and a broadening of the extrusion zone with concanavalin A introduced into jejunal blind loops. In addition, there are large amounts of cell debris and mucus in the lumen after exposure to the lectins. However, the binding of concanavalin A to brush border membranes, and all damaging effects following this binding, can be inhibited by high concentrations of D-mannose. As can be expected, pre-treatment of the tissue with neuraminidase liberates more free terminal mannose residues in the membrane glycoproteins with a consequent increase in the amounts of concanavalin A bound. It is, therefore, not surprising that the *in vivo* effectiveness of the lectin is enhanced by neuraminidase treatment. When concanavalin A is perfused in combination with neuraminidase *in vivo* for 10 days, the weight loss of the rats is more serious than in controls which have been treated with concanavalin A alone. In addition, the morphological changes in brush border membranes are also the most striking in this group of rats. Overall, the damaging effects of the lectin-induced hyper-regenerative adaptive changes strongly resemble those found in coeliac disease (Lorenz-Meyer *et al.*, 1985).

In neonatal animals, lectins may also damage the gastric mucosa. Thus, the damage caused by intrapharyngeally infused concanavalin A in neonatal guinea pigs is more serious in the gastric than in the small intestinal mucosa (Weaver & Bailey, 1987). Morphometric analyses of the jejunum have revealed a slight decrease in villus height while crypts significantly lengthened, with a 4.7-fold increase in jejunal crypt cell production rates. The integrity of the mucosa is increasingly compromised by exposure to concanavalin A and the presence of the lectin can be demonstrated in the crypt cells within 60 minutes. These results indicate that the neonatal gastrointestinal tract is more susceptible to the damaging effects of food lectins than the mature one (Weaver & Bailey, 1987).

The reactivity of the small intestine to food proteins and lectins differs significantly in newborn and adult animals. The adult small intestinal mucosa can bind more concanavalin A than the neonate; in strong contrast to the non-specific binding of milk proteins which, is more extensive in newborn animals than in adults. It is, therefore, difficult to see the relevance of lectin studies to the exploration of the mechanism of intolerance to milk (food) proteins in newborn animals or, indeed, in coeliacs (Stern & Gellerman, 1988; Stern *et al.*, 1988).

A comparison of the effects of concanavalin A and PHA, the two most intensively studied lectins, shows that the two lectins induce similar and harmful lesions in the small intestine despite their different sugar specificities. It is remarkable that, despite binding to different parts of the villus epithelium, both lectins cause serious interference with digestion and absorption in the gut by a similar mechanism although the fine details of their reactions with brush border cells is unknown.

Lectins from Phaseoleae *(other than* vulgaris). Several members of the *Phaseolus* genus contain lectins similar to PHA (Pusztai *et al.*, 1983a; Grant *et al.*, 1983). Although some of these beans are used as food in a number of Third World countries, their lectin-related toxicity severely restricts the use of them (Pusztai, 1985, 1986a).

Unfortunately, very few of these seeds have been studied in any detail and, in contrast to the extensive studies with PHA, no comparable effort has been spent on studying the biochemical, nutritional and immuno-chemical properties of the constituent lectins in any of the seeds within the members of this genus. For example, it is known that runner beans, *Phaseolus coccineus*, are nutritionally toxic (de Muelenaere, 1965) and contain a well-defined lectin (Angelisova & Haskovec, 1978), but no connection between the seed lectin and toxic effects has been demonstrated. Similarly, although a crude lectin preparation from lima bean, *Phaseolus lunatus*, restricts the growth of rats (Manage, Joshi & Sohonie, 1972), it is not known if this toxic effect is due to the lectin in the seed or to some other component.

Extracts of tepary bean, *Phaseolus acutifolius* cv. *latifolius*, have severe toxic effects on rats which include the widespread destruction of brush border microvilli, disruption of the endoplasmic reticulum and swelling of mitochondria in epithelial cells of the rat small intestine

(Sotelo *et al.*, 1980, 1983). By analogy with PHA, such damaging effects of tepary bean extracts are strongly indicative of lectin-induced lesions. Both erythro- and lymphoagglutinins, which are very similar to PHA isolectins, have been isolated from the seeds and characterized (Pusztai, Watt & Stewart, 1987). Nevertheless, the effects of pure tepary bean isolectins on the rat small intestine have not been tested.

Gluten lectins–wheatgerm agglutinin–coeliac disease. Coeliac disease is a characteristic disease of the small intestine of humans. It is a primary intolerance of the human gastrointestinal tract to wheat gluten and/or gliadin peptides which occurs in genetically predisposed individuals (Auricchio *et al.*, 1985). It is a disease for which there are no ready-made animal models. As mentioned previously (p. 168), there have been attempts to use concanavalin A as a model to study the human disease in animals but most attempts have been only partially successful (see above).

Although the aetiology of the disease is complex and not well understood, the symptoms are well known and include malabsorption, diarrhoea and atrophy of the small intestinal mucosa. It is certain that in susceptible individuals the major histocompatibility complex is involved. Indeed, during the development of the disease the involvement of the complex is expressed through a cell-mediated immune hypersensitivity reaction to gluten proteins and peptides. However, one of the primary causes for the disease may be due to the binding of cereal lectins to immature surface determinants of coeliac enterocytes. The effects of the bound lectins on the small intestine are probably twofold. In the first place, the lectins may damage the brush border in a way similar to the injury caused by lectins, such as PHA or concanavalin A. However, it is also possible that it is not the damage due to the initial binding which is the main cause of the disease, but that because of the attachment of the lectin to the mucosa, the stimulation of the gut associated lymphoid tissue is prolonged. Thus, the extent of the stimulation of the gut immune system by an antigen which is relatively stable against gut proteolysis and binds positively to immunocompetent lymphocytes is expected to be higher than that by other food antigens which do not bind to the gut mucosa.

The identity of the lectin(s) in gluten is controversial. There is a large body of evidence to suggest that one or several components in the gluten and/or gliadin fractions of wheat endosperm behave *in vitro* as a polymannose specific lectin (Köttgen *et al.*, 1982, 1983; Concon, Newburg & Eades, 1983; Auricchio *et al.*, 1984). However, *in vivo*, the binding of gluten lectins to the human small intestinal epithelium has not yet been established unequivocally. Nevertheless, gluten lectins bind to the small intestine of 17 to 18 day-old rat foetuses and they can also inhibit the full development of the immature foetal cells to fully mature ones. In keeping with the proposed theory, the same gluten lectins do not bind to the fully differentiated and mature cells in newborn rats (Auricchio *et al.*, 1982). These results are in accord with experimental findings in human coeliacs, whose small intestine contains relatively large numbers of immature epithelial cells. It is suggested that the conversion of the high-mannose type oligosaccharide chains to mature and complex carbohydrate-containing glycosidic groups in the membrane glycoproteins of the small intestinal enterocytes is blocked, and that these mannose residues serve as the main receptor sites for gluten lectin binding in coeliacs.

It has also been suggested recently that it is not the gluten lectins which bind to the small intestinal epithelium but that the membranes of epithelial cells react with the wheatgerm agglutinin which usually contaminates gluten preparations (Kolberg & Sollid, 1985). Higher antibody titres to wheatgerm agglutinin found in coeliacs compared to those in healthy controls appears to support the involvement of wheatgerm agglutinin in the disease (Sollid *et al.*, 1986). Unfortunately, antibody titres are a somewhat unreliable guide to the true cause for the development of the coeliac syndrome, as usually in the same individuals, the antibody response to other food proteins is also increased. Wheatgerm agglutinin is known to increase the permeability of the small intestine (Lorenzsonn & Olsen, 1982) and this lectin may make a definite contribution to the general leakiness of the small intestine of coeliac individuals exposed to wheat gluten diets. As a result, there is likely to be an increased absorption of potentially cytotoxic intact proteins of cereal diets through the wall of the small intestine into systemic circulation. The ensuing interference with both the local (gut) and the body's immune system may lead to the full development of the chronic coeliac syndrome. Whether either or both types of lectins is responsible for the initial binding, the net effect on immature enterocytes is a serious damage to their membranes, with consequent reduction in their normal absorptive function. This initial cellular damage may then be followed by cell-mediated and other immune hypersensitivity responses characteristic of the fully developed coeliac disease.

As wheatgerm agglutinin may increase the leakiness of the small intestine it has been used frequently as a possible model for studies into the mechanism of intestinal allergic reactions. While the lectin instilled into ligated segments of the distal small intestine of rats decreases its permeability to simultaneously introduced small molecular weight dextrans, the uptake of the larger ($M_r = 3000$) polymers increases significantly (Sjölander, Magnusson & Latkovic, 1984). This is in accordance with what is generally found in patients with coeliac disease. One explanation for this paradoxical behaviour is that, while the small molecules diffuse through pores in the membranes, the larger molecules can penetrate into the epithelium through paracellular routes or the cell extrusion zones. Both routes appear to be more available as a result of selective permeability changes caused by wheatgerm agglutinin. As the situation is reversed when concanavalin A is used instead of wheatgerm agglutinin, the small intestine appears to respond to the two lectins in a different way. It is possible that, owing to the mucotractive effect (Freed, 1979) of concanavalin A, the mucus released by mast-cell mediated hypersecretion from the small intestinal epithelium may form an additional barrier to the absorption of larger molecules. Accordingly, concanavalin A may cause a low grade food allergy, while wheatgerm agglutinin does not. However, it is very difficult to relate the results of these permeability studies carried out with relatively small marker molecules as true indicators of the behaviour of the small intestinal epithelium in coeliac disease or even in allergic reactions (Sjölander *et al.*, 1984).

The more recent studies by Sjölander and her colleagues appear to be more relevant as models for the pathogenesis of coeliac disease (Sjölander *et al.*, 1986). Both wheatgerm agglutinin and concanavalin A affect the morphology of the rat small intestine. As expected from their binding specificities, both lectins bind to the enterocytes at the base of the villi rather than at the top. The resulting damaging changes include disarrangement of the cytoskeleton, shortening of the villi and, most importantly, increased endocytosis of the lectins and other macromolecules. The increased rate of membrane turnover may be the result of the interference with cellular metabolism by the endocytosed lectins. These results may, therefore, bear a resemblance to the morphological changes occurring in the small intestine of coeliacs (Sjölander *et al.*, 1986).

It is apparent from the intracellular events which follow the binding of wheatgerm agglutinin to intestine 407 cells in cell cultures that wheatgerm agglutinin is a strong extracellular growth signal (Sjölander, 1988). Direct stimulation of the cells with this lectin leads to an immediate rise in cytosolic free calcium concentration; a part of which is derived from the mobilization of intracellular stores and is probably mediated via the simultaneous increase in inositol triphosphate formation. However, the bulk of the increased calcium in the cytosol is due to the influx of extracellular Ca^{2+} from the medium. As the wheatgerm agglutinin-based signal transduction mechanism is only slightly sensitive towards pertussis toxin, the G protein which is involved in coupling the external lectin signal to intracellular metabolic processes appears to be different from those described for other signal transduction processes. These studies have, therefore, reinforced the concept of the interactions between mitogenic lectin signals and the consequent cellular responses in the small intestinal epithelium, first explored in detail with the PHA-induced and polyamine-dependent hyperplastic growth of the small intestine.

Miscellaneous lectins. Psophocarpus tetragonolobus, winged bean, is one of the most promising, recently introduced tropical legumes, with great potential as food for both humans and animals. Practically all parts of this plant, including seeds, tubers, stems and leaves, are edible. Unfortunately its use is limited by the presence of a number of deleterious constituents, particularly when fed in the raw state. The most important of the anti-nutritive factors are two lectins and the trypsin inhibitors. Thus, before the plant, and particularly the seeds, can be used safely for feeding, there is a need for detoxification by proper heat treatment.

Detailed fractionation studies carried out in conjunction with nutritional testing have revealed that the main factors responsible for most of the toxic effects are the constituent seed lectins (Higuchi, Suga & Iwai, 1983; Higuchi, Tsuchiya & Iwai, 1984). Winged bean lectins show the usual high resistance to gut proteolysis, generally characteristic for lectins. In all other respects too, the reaction mechanism of the interaction between the winged bean lectins and the cells of the small intestine is very similar to that previously explored with others, such as PHA or concanavalin A. Winged bean lectins bind avidly to brush border cells *in vivo* and the ensuing serious disruption of the highly ordered structure of the microvillus membrane of epithelial cells impairs the digestive and absorptive functions of the small intestine and also hinders the normal maturation process and replacement of brush border cells (Higuchi *et al.*, 1984; Kimura *et al.*, 1986).

Two legumes grown extensively and consumed widely in India, horse gram (*Dolichos biflorus*) and hyacinth bean (*Dolichos lablab*) also contain lectins. By analogy with the toxic effect of PHA or concanavalin A, the lectin constituents of these Indian legume seeds are also implicated in the anti-nutritive effects. However, when tested experimentally, partially purified preparations of these Indian legume lectins are less toxic for rats than the raw seed beans (Manage *et al.*, 1972).

Peas (*Pisum sativum*) or lentils (*Lens culinaris*) are widely cultivated plants in temperate climates; their seeds are widely consumed by both humans and animals and are the least toxic of all the presently grown legumes. The nutritional utilization of these legume seeds is only slightly improved by heat treatments, suggesting the presence of relatively small amounts of anti-nutritive substances. It is, therefore, somewhat surprising to find that the incorporation into diets of isolated or partially purified preparations of either of the two lectins apparently reduces the activities of a number of key brush border hydrolytic enzymes and retards the growth of rats (Jindal, Soni & Singh, 1982). The transport of sugars and amino acids by brush border membrane vesicles is also inhibited by the presence of pea or lentil lectins (Dhaunsi *et al.*, 1985). However, the stability against proteolysis of the pea lectin is considerably less than that of other lectins, such as PHA or concanavalin A. Thus, only about 15% of radiolabelled pea lectins, initially administered, escapes digestion during passage through the gastrointestinal tract to the colon unaltered. Despite the presence of functional pea lectin in the gut, its binding to the small intestinal mucosa was slight (Aubry & Boucrot, 1986). This was confirmed recently for both pea and *Vicia faba* lectins (Fig. 6.21; Pusztai *et al.*, 1990a). Consequently, their effect on the morphology of the small intestine is also slight (Fig. 6.22). Thus, from the viewpoint of *in vivo* nutritional effects, the *in vitro* reduction of brush border hydrolytic enzyme activities has little significance.

One of the most interesting lectins is that obtained from the tomato (*Lycopersicon esculentum*) fruit. Although the tomato lectin included in diets shows a high resistance to proteolytic breakdown in the alimentary canal and binds to intestinal villi, only small amounts of it can traverse the gut wall and become systemically absorbed (Kilpatrick *et al.*, 1985). Moreover, despite binding to the brush border epithelium, tomato lectin has practically no destructive effects on the normal morphology of the gut and the tomato lectin is, therefore, not toxic. Taking into consideration the specificity of the lectin for *N*-acetyl-D-glucosamine oligomers and the large amounts of tomatoes that are consumed by humans, it is rather curious that there are no reports of any adverse or other effects of this lectin on human coeliacs. Indeed, tomato lectin ought to bind to the same regions of the small intestinal epithelium as wheatgerm agglutinin. These sites may also be close to the polymannose receptor sites of putative gluten lectins. Accordingly, the consumption of tomato lectin or tomatoes may have beneficial effects on coeliacs by blocking small intestinal receptor sites for gluten lectins and/or wheatgerm agglutinin (Kilpatrick *et al.*, 1985).

Lectins are also present, sometimes in relatively large amounts, in plants which are not used as foods for humans or feeds for economically

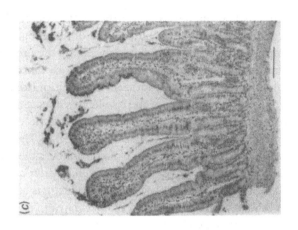

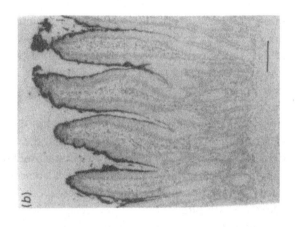

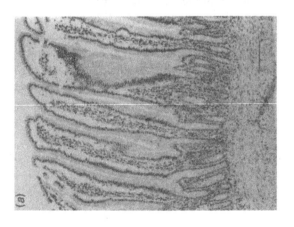

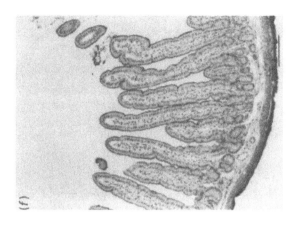

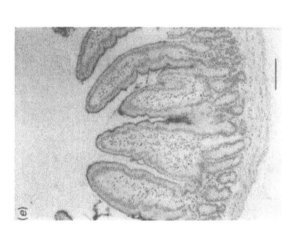

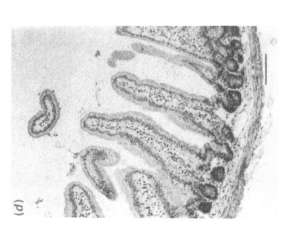

Fig. 6.21. Specific immunoreactive PAP-staining of sections obtained from the jejunum of rats fed on the different lectin diets for 10 days. Formalin-fixed sections after trypsin treatment were reacted with the appropriate monospecific rabbit anti-lectin antibodies. Second antibody treatment (PAP) was followed by staining the sections with 3,3′- diaminobenzidine and counterstaining by haematoxylin. Bar represents 100 μm. (a) VfL (*Vicia faba*); (b) PHA; (c) SBA (soyabean, *Glycine max*); (d) SNA-I; (e) SNA-II (*Sambucus nigra*) and (f) GNA (*Galanthus nivalis*). (From Pusztai et al., 1990a; by permission of *Digestion*.)

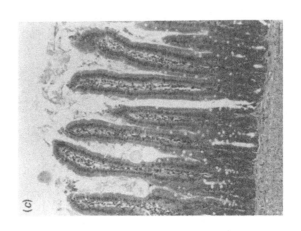

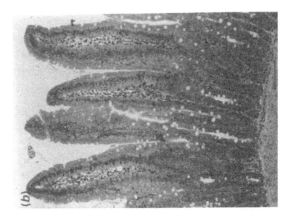

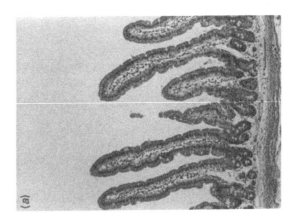

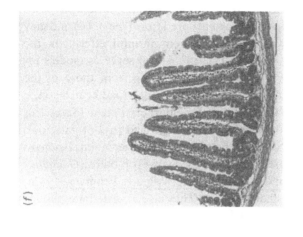

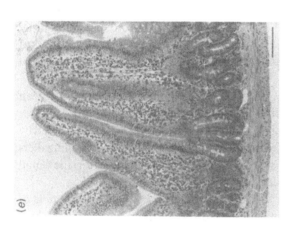

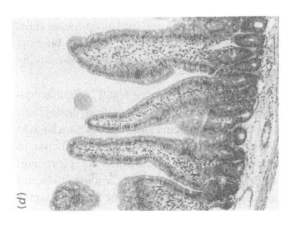

Fig. 6.22. Light micrographs of formalin-fixed and haematoxylin-eosin-stained sections (3 μm) of the jejunum of rats fed on different lectin diets for 10 days. The following sections were examined: (a) control (lactalbumin-fed); (b) PHA; (c) SBA (*Glycine max*); (d) SNA-I; (e) SNA-II (*Sambucus nigra*) and (f) GNA (*Galanthus nivalis*). Sections obtained from rats fed on broad bean (*Vicia faba*) lectin were similar to those of control rats and were not given. (From Pusztai *et al.*, 1990*a*; by permission of *Digestion*.)

important domestic animals. Such lectins, particularly those from bulbs or barks, may be defensive agents against predators (see p. 60) and may fulfil their biological function through their harmful effects on the alimentary tract of the predators. These may also serve as models for lectin–gut interactions of varying severity ranging from those of the nutritionally toxic PHA and the relatively harmless broad bean lectin.

The two lectins from the bark of the elderberry plant (*Sambucus nigra*), SNA-I (specific for *N*-acetylneuraminic acid [α-2-6] galactose; Shibuya *et al.*, 1987) and SNA-II (specific for galactose; Kaku, Peumans & Goldstein, 1990) and the lectin from the snowdrop bulbs (*Galanthus nivalis*; specific for D-mannose; van Damme, Allen & Peumans, 1987*a*) survive the passage in the small intestine. They also have definite, though variable, effects on the morphology and growth of both the small intestine and the pancreas (Figs 6.15–6.22). Furthermore, feeding rats for 10 days on semi-synthetic diets containing 0.7% of the individual lectins depresses the growth of the rats to a variable extent, dependent on the lectin fed. Accordingly, there is a correlation between the effectiveness of the lectins as growth promoters for the small intestine and their inhibition of body growth.

In addition to the toxic PHA and the slightly less effective soyabean agglutinin, SNA-II also binds extensively to the mucosa after acute or continuous dietary exposure to the lectin. As all three are endocytosed by epithelial cells, particularly after feeding rats with lectin-containing diets for 10 days, these lectins are powerful growth factors for the small intestine by inducing hyperplastic growth of the mucosa.

The binding of SNA-I, which is moderate to slight at the beginning of the exposure to the lectin, is reduced on continuous exposure. Thus, SNA-I is only a moderately effective growth factor for the small bowel. The same applies to GNA, the binding of which is slight at the start but increases on extended periods of feeding. In contrast, the non-toxic *Vicia faba* lectin is neither bound to, nor is endocytosed by, the epithelial cells. This confirms the existence of a firm correlation between binding/ endocytosis (i.e. lectin function) and the *in vivo* gut growth-promoting activity (Pusztai *et al.*, 1990*a*).

Finally, as glycoproteins may also bind to the small intestinal epithelium and this may be specifically inhibited by carbohydrates, our concepts of interactions between food proteins and gut cells may have to be revised. Thus, the recent demonstration of a significant reduction in the binding of fucose-free lactoferrin to small intestinal receptors indicates that lectin–glycan interactions are not necessarily confined to lectins from the food and mucosal receptors in the gut. The opposite process may

Table 6.9. *Common features of toxic (anti-nutritive) effects in lectin–gut interactions.*

1. High degree of resistance to gut proteolysis

2. Binding to brush border cells; damage to microvillus membrane; shedding of cells; reduction in the absorptive capacity of the small intestine

3. Increased endocytosis; induction of hyperplastic growth of the small intestine; increased turnover of epithelial cells

4. Interference with the immune system; hypersensitivity reactions

5. Interference with the microbial ecology of the gut; selective overgrowth

6. Direct and indirect effects (hormones, etc) on systemic metabolism

also occur, since food glycoconjugates can also be selectively and specifically bound by gut endogenous lectins (Davidson & Lonnerdal, 1988).

In addition to those discussed above, other lectins may also have similar, but so far unexplored, effects on cells and membranes of the small intestinal epithelium. However, for the most intensively studied lectins, some common features of the mechanism of their interactions with gut cells are beginning to emerge (Table 6.9).

Transport of lectins through the membranes of the adult gut – a potential system for the oral delivery of drugs

As has been discussed in previous sections, most of our understanding of the effects of lectins on the gut and their consequences is based on experimentation with a few lectins, such as PHA, concanavalin A, wheatgerm agglutinin or soyabean lectin. It is of particular importance to establish if there is a connection between lectin function and systemic absorption and, if there is, what features of the lectin molecules determine the uptake process. Although it appears from oral immunization studies that systemic absorption of small amounts of functional lectins is a commonly occurring process (de Aizpurua & Russel-Jones, 1988), it is not clear how far such a process can be used to deliver proteins, polypeptide hormones and other drugs reliably and in sufficient amounts into the body through the oral route.

The systemic absorption of lectins, when it occurs, affects the metabolism of the body. From such a viewpoint, the internalized lectin molecules may be regarded as polypeptide or protein-like hormonal drugs. For example, the modulation of systemic insulin, glucagon and, possibly, other hormone levels of rats fed on PHA and the ensuing changes in general metabolism, are clear-cut drug effects. Therefore a systematic exploration of the process of the lectin-mediated endocytosis may point the way to potential medical applications of this fundamental biochemical and cell biological reaction.

Most microbial or plant lectins and/or toxins interact with cells of the small intestine. Some of the consequences of these interactions have already been discussed in detail in previous sections, although the lectins described so far belong to the first group of the two main classes of lectins/toxins found in Nature. These are made up exclusively of lectin subunits. However, lectins belonging to the second group of naturally occurring lectins also bind to epithelial cells of the small intestine and are taken up by endocytosis. Some of these lectins, and particularly those of microbial origin, are regarded first and foremost as toxins rather than as lectins as their effects are usually more damaging overall. Nevertheless, despite their high toxicity, these substances also possess clear-cut features characteristic for lectins.

The structure of this second type of plant or microbial toxins/lectins is quite distinct from the first type which contains only lectin subunits. The main difference is that the second type of toxins/lectins have a basic A–B subunit structure, where the A subunit is a toxin and only the B subunit is a lectin. The intracellular absorption of these molecules by endocytosis is strictly dependent on the B subunit. Toxic effects occur after the A chain is delivered into the cell and becomes dissociated from the A–B molecule (Waksman *et al.*, 1980). Although there are differences in the transfer of the various A–B toxins through the plasma membrane of cells, a number of basic features appears to be common for all. Thus, after the B subunit binds to the cell membrane, its hydrophobic part is inserted as a channel into the membrane (Blaustein *et al.*, 1987). However, in some instances, such as in the diphtheria toxin translocation, the A fragment may also take part in this process. The toxic A subunit, still attached by a disulphide bridge to the lectin and bound to the acidic endosomal organelles, enters the cytoplasmic face of the membrane through this channel. After delivery into the cell, the diphtheria toxin changes its conformation at low pH and penetrates the membrane of this acidic organelle to reach the cytoplasm, where the A subunit is released from the intact toxin as a result of reduction by cytoplasmic glutathione and/or proteolytic modification (Zhao & London, 1988; Gonzalez & Wisnieski, 1988; Dumont & Richards, 1988). Some of the other toxins, such as ricin, abrin and a number of immunotoxins, do not require transfer to a low pH compartment for translocation through the membrane or a conformational change at low pH (Olsnes *et al.*, 1988). The free toxin then reacts with the subcellular organelles of the cell (van Deurs *et al.*, 1988), such as the Golgi apparatus, or becomes sequestered into lysosomal structures or even reaches the nucleus. A large part of the toxin may be rendered harmless by proteolytic digestion in the lysosomal compartments. Although the processes of translocation of the diphtheria toxin into the plasma membrane and, subsequently, from the endosomic particle into the cytosol are relatively inefficient and a part of the toxin may also be broken down, the presence of even small amounts of functional A subunits will irreversibly damage the protein synthetic apparatus of the cell. The A fragment is an enzyme which, by ADP-ribosylating a unique amino acid residue, diphthamide, in elongation factor 2, inactivates it and, thereby, blocks protein synthesis in the cell (van Ness, Howard & Bodley, 1980). However, small amounts of the toxin may also be removed from the cells by the reverse process of exocytosis and reach the serosal side of the epithelium. On the other hand, only a part of the B fragment is protected from proteolytic breakdown and this probably remains inserted into the membrane (Moskaug, Sandvig & Olsnes, 1988).

Thus, despite the obvious differences between the various A–B toxins as regards their translocation through membranes, the presence of the B subunit appears to be of paramount importance for exerting toxic effects inside the cells. Apparently, it is not only necessary for the initial attachment of the toxin/lectin to the cell membrane, but also for forming channels or tunnels through it (Blaustein *et al.*, 1987). It is now well known that toxic A subunits, without attached lectin chains, are widely distributed in plants (Stirpe & Barbieri, 1986). However, these ribosome-inactivating peptides, unlike the A–B toxins, are active only in cell lysates as they are not transported through the membranes of intact cells. In contrast, the B subunits are able to traverse membranes, but they are not lethally toxic without the presence of the A subunits.

Absorption of toxins via the tunnel-forming lectin subunits occurs widely in both plants and microorganisms. Of the bacterial toxins several examples have been studied extensively. These include, amongst others, the diphtheria toxin, *Pseudomonas aeruginosa* exotoxin A (Gill, 1978), the heat-labile *Escherichia coli* enterotoxin (Dallas & Falkow, 1980) and the *Clostridium perfringens* enterotoxin (McDonel, 1980). The most studied A–B toxins from plants are ricin, abrin, modeccin, volkensin and the mistletoe lectin, viscumin (Pusztai, 1989*c*).

It is clear from a comparison of their effectiveness after injection intravenously or having been given orally that A–B toxins, such as ricin, are far less toxic given intragastrically than parenterally (Ishiguro *et al.*,

1983, 1984). In fact, the reaction of ricin with gut epithelial cells resembles that of PHA or concanavalin A and ricin binds to gut cells through its B subunit. Its effects on the morphology and functioning of the small intestine are also somewhat similar to those obtained with food lectins. Furthermore, the extent of its binding to the intestinal epithelium is dependent on postnatal maturation and the attachment of the toxin to epithelial membranes is much higher in suckling than in adult rabbits (Olson *et al.*, 1985). In the adult rat, binding is followed by time- and dose-dependent morphological changes, including villus atrophy, crypt elongation, degeneration of the epithelium, fusion of intervillous epithelia, infiltration of neutrophils and eosinophils and dissociation between the epithelium and the lamina propria of the small intestine when ricin is administered in the dose range of 1 to 60 mg/kg. The changes are most obvious in the jejunum and at the highest doses (Sekine *et al.*, 1986). Although the rats die eventually, the doses of ricin required for this are about two to three orders of magnitude higher than the lethal dose on parenteral administration. These results confirm the results of early experiments by Ehrlich, who has managed to immunize animals systemically against the toxic effects of parenterally introduced ricin by repeatedly feeding them on low concentrations of ricin as a part of their diets.

Clearly, initially the binding and all morphological changes observed with ricin are lectin effects and most of these changes can also be observed with the separated B-subunits. It is also clear that the eventual lethal effects of the orally introduced A–B toxin ricin are due to the A chain which originates from the small amounts of toxin which traverse the gut wall and become systemically absorbed. Therefore, it may be possible to exploit the reactivity of the B-subunit with the gut epithelium for practical delivery of drugs by the oral route. Thus, if in place of the toxic A-subunit, drug or hormone polypeptides or proteins are attached to the B-subunit or other similar lectin subunits, they may cross into the systemic circulation together with the lectin (Pusztai, 1989*c*). Similarly, the well-known lectin-mediated agglutination and fusion of glycolipid-containing phospholipid vesicles may also be exploited for oral drug delivery (Hampton, Holz & Goldstein, 1980; Slama & Rando, 1980; Alecio & Rando, 1982; Hoekstra & Duzgunes, 1986). When these liposomes are targeted to bind to the small intestinal epithelial cell membranes through the lectin part, the difficulties encountered with the lack of absorption through the gut wall of liposome-encapsulated drugs may be successfully overcome (Chiang & Weiner, 1987). Indeed, lectin-mediated targeting of liposomes to model surfaces coated with glycophorin (blood group B) has already been achieved with a relatively high efficiency (Hutchinson &

Jones, 1988). As most brush border membrane glycoproteins carry carbohydrate determinants appropriate for the blood group of the animals, correct targeting may be attained by using lectins of the right specificity for the particular blood group.

Concanavalin A, in addition to its specificity for D-mannose/glucose, also recognizes myo-inositol (Wassef *et al.*, 1985). Thus, with concanavalin A, the fusion of liposomes with epithelial membranes may occur without the presence of glycolipids in the liposomes. Moreover, concanavalin A and other lectins have a strong affinity for hydrophobic sites (Maliarik & Goldstein, 1988; Maliarik *et al.*, 1989) in cell membranes. This may be high enough to aid the fusion of liposomes with the lipid bilayers of small intestinal epithelial cell membranes and to ensure delivery of a bound drug (Pusztai, 1989c).

As there is a possibility of disulphide-bond formation between the A-subunit and potential polypeptide ligands, the tunnel-forming B subunits of the ricin type may be the most obvious choice for drug delivery trials. Such a complex may also mimic the behaviour during membrane transfer of the naturally occurring A–B toxins if a part of the endocytosed A–B or B type lectins may be removed from the gut epithelial cells by the reverse process of exocytosis through the basolateral membrane and delivered in sufficient quantities into circulating blood or the lymphatics. This route of drug delivery may also be augmented by the transport of macromolecules bound to lectins, such as PHA (Fig. 6.10) through M cells of Peyer's patches of small intestinal epithelium. Accordingly, the successful oral administration of polypeptide drugs may become a reality.

Based on the increased transit time of oral formulations in the presence of lectins, the efficiency of drug delivery from the small intestine into blood circulation may also be increased. The simultaneous application of a lectin with an orally given drug, even if it is not conjugated to the drug, slows down its passage through the alimentary tract (Woodley & Naisbett, 1988), and, because of the binding of the lectin to the gut mucosa, the transit time of drugs conjugated to a lectin may be even more extended. As tomato lectin is non-toxic, resistant to proteolysis in the small intestine and binds avidly to the mucosa, it may be a choice candidate for such an application of controlled release of orally introduced drugs (Woodley & Naisbett, 1988).

Interactions with nerve cells and transneural transport of lectins
Mature cells of the nervous system are metabolically active even though they do not undergo cell division. Like all other eukaryotic cells, the membrane glycoconjugates of neurons can bind lectins, and various

lectins are frequently used as reagents to localize carbohydrate-containing sites on membranes of nerve cells (Biltiger & Schanebli, 1974). Furthermore, several lectins are axonally transported (Dumas, Schwab & Thoenen, 1979; Fabian & Coulter, 1985).

Wheatgerm agglutinin. The axonal transport of wheatgerm agglutinin has been particularly widely used for anatomical studies of neuronal connections (Trojanowski, 1983) as it is well known to be transported in both anterograde and retrograde directions. In addition, wheatgerm agglutinin is also transported transneuronally from axonal endings into adjacent neurons. Thus, wheatgerm agglutinin was found by immunocytochemical methods in the neuronal somata in the vicinity of afferent terminals after anterograde transport in retinotectal and trigeminal afferents (Ruda & Coulter, 1982). Similar transneural transport of this lectin also occurs in other systems (Borges & Sidman, 1982) and after retrograde transport (Harrison *et al.*, 1984).

The existence of the transneuronal transfer of wheatgerm agglutinin is a clear demonstration of the occurrence of exchange of large protein molecules between nerve cells. Signal transduction occurs widely between different neurons after axonal delivery and the transference of glycoproteins into muscle neurons by nerve stimulation (Younkin *et al.*, 1978). Indeed, the very same glycoproteins have also been implicated in the endocytosis and transport of lectins (Brunngraber, 1969; Edelman, 1983).

The axonal and transneuronal transport of lectins appears to be dependent on their carbohydrate specificities (Fabian & Coulter, 1985). Immunocytochemical staining methods show that wheatgerm agglutinin, which recognizes N-acetyl-D-glucosamine and its oligomers or the three mannose-binding lectins; concanavalin A; pea and lentil lectins, are transported retrogradely to the facial nucleus after injection into facial muscles. In contrast, all four lectins are transported anterogradely to the optic tectum after injection into the vitreous body. Soyabean lectin, with specificity for D-galactose and/or N-acetyl-D-galactosamine, undergoes only slight retrograde transport, while neither the agglutinins of peanut or *Ulex europaeus* are carried by axonal transport. Moreover, by immunocytochemical localization, those lectins which can move axonally in an anterograde direction are seen in the neuronal somata at the afferent terminals. None of the other lectins is transported transneuronally. These studies therefore support the existence of protein–signal exchange between different neurons and the mediation of trophic or other stimuli to target neurons (Fabian & Coulter, 1985).

The uptake and the transport of wheatgerm agglutinin has also been studied by the intravitreal injection of iodinated wheatgerm agglutinin (^{125}I-labelled) instead of the more static anatomical approach used by Fabian & Coulter (1985). The labelled lectin, which is not toxic at the concentrations used, is taken up by chick retinal ganglion cells and transported anterogradely to nerve terminals in the optic tectum where its accumulation can be measured quantitatively. Co-injection of un-labelled wheatgerm agglutinin with the labelled lectin reduces the axonal transport of the radioactive label, while lectins from soyabean or *Ulex europaeus* are without such an effect. Thus, the uptake and subsequent anterograde transport of wheatgerm agglutinin is a selective process and depends on the existence of a number of extra- and intracellular binding sites for the lectin. It has been suggested that a portion of the endocytosed wheatgerm agglutinin becomes transported via a pathway involving the smooth endoplasmic reticulum in an anterograde direction and is then released at retinal ganglion cell terminals (Margolis & LaVail, 1984).

Wheatgerm agglutinin has been a popular choice of reagent for the mapping of axonal and transneuronal pathways because of its slight toxicity and high reactivity with extra- and intracellular receptors containing N-acetyl-D-glucosamine and/or sialic acid. Derivatives, such as horseradish peroxidase conjugated- or fluorescein thiocyanate-labelled wheatgerm agglutinin, have been used in some of the more dynamic studies to aid tracing these pathways. However, it is by no means certain that the conjugated lectin probes behave in the same way as the unconjugated lectin. Indeed, anterograde transport rates in the retinotectal projection of posthatched chicken are different for the native lectin (168 mm/day) and the horse peroxidase-conjugated wheatgerm agglutinin (345 mm/day). Retrograde transport rates are in the range of 150–180 mm/day. Moreover, the fluorescein isothiocyanate-labelled lectin is transported retrogradely but not anterogradely, hence, the different conjugates of wheatgerm agglutinin or possibly even the native lectin, are transported in different components of the axonal transport (Crossland, 1985).

In a different approach, horseradish peroxidase-conjugated wheat-germ agglutinin has been injected into the submandibular gland of rats. Both lectin-conjugated and unconjugated horseradish peroxidase immunobands were revealed by immunoblot studies with anti-horseradish peroxidase antibodies in homogenates of the injected glands and superior cervical ganglia, which contain neurons retrogradely labelled with the conjugated lectin. For reasons which are not clear at present, with antiserum to wheatgerm agglutinin, no immunobands have been detected in the superior ganglia homogenates. In contrast, monomeric wheatgerm agglutinin is present in the submandibular gland for about 24 hours after the injection of the conjugated lectin. Thus, the intraneuronal endocytotic and transport pathways and degradation of proteins can be mapped from the molecular composition of the lectin conjugates before and after retrograde transport (Schmidt & Trojanowski, 1985).

Wheatgerm agglutinin conjugated with apo-horseradish peroxidase–gold provides an even more sensitive method of retrograde tracing of the projections of the neurons of the central nervous system. This tracer can be used for both light- and electromicroscopy and in single and double-label studies and, as it uses an enzymatically inactive form of horseradish peroxidase, the tissues do not require pre-treatment. Moreover, at the electron microscope level, the gold label can be easily distinguished from the immunoreaction product (Basbaum & Menterey, 1987).

It appears that, for consistent transneuronal labelling, neonates require the injection of much higher amounts of horseradish peroxidase-conjugated wheatgerm agglutinin into the visual system than adult rats (Haya, 1988). However, at this increased level, transneuronal labelling can be observed at all postnatal ages (Itaya, 1988). The progression of the lectin conjugate along the geniculocortical pathway with age parallels the growth and development of the geniculate axons although the reasons for this age-related difference are not clear. The lectin first appears in the terminal layers of the visual cortex 7 days after birth, a few days after the first observation of thalamocortical synapses (3–4 days). It is possible that the lectin may be located in the growing fibres and growth cones, and that the conjugated lectin-containing vesicles may be preferentially targeted for specific avenues of intracellular transport. Similar targeting of the lectin conjugate may also hold for the adult rats. Transport via the lysosomes or the Golgi pathway may thus be separate and the passage through the *trans*-Golgi cisternae may be characteristic for glycoproteins destined for fast axonal transport (Hammerschlag *et al.*, 1982).

The growth of nerve fibres may be regulated by receptors on the surface of growth cones and this could be affected by their lectin-induced redistribution (Campenot, 1977). Two lectins, wheatgerm agglutinin and concanavalin A, with different sugar specificities, have almost opposite effects on growing nerve fibres in culture. Thus, wheatgerm agglutinin reacts with growth cones and reversibly inhibits the elongation of the nerve fibre, whilst concanavalin A stimulates the initiation of nerve fibres. These differential effects of lectins on fibre growth are specific and may mimic the differences in the actions of various growth factors which determine the direction and the rates of nerve fibre growth *in vivo* (DeGeorge & Carbonetto, 1986).

Phaseolus vulgaris *lectin*. PHA has also been used to examine pathways of the central nervous system although somewhat less frequently. For example, an anterograde tracing technique has been developed based on injection, uptake, transport and subsequent immunocytochemical visualization of L_4-PHA. With this lectin, only slight retrograde transport occurs and neuronal connections can be traced with high degree of accuracy (Wouterlood, Bol & Steinbush, 1987). Similarly, the projections of the medial preoptic nucleus in the rat have been mapped with *Phaseolus vulgaris* leucoagglutinin from an anterograde tract-tracing study (Simerly & Swanson, 1988). The use of PHA has some advantages as an anterograde tracer in the central nervous system, which includes a Golgi-like and discrete labelling of axons and a short distance (less than 20 mm from the injection site) labelling of the fibres of passage. The stability of the lectin for months is particularly important as its transport is very slow, 4–6 mm/day.

PHA-L_4 is also a useful anterograde tracer in the peripheral nervous system; it is transported anterogradely into the spinal cord after intracellular injection into dorsal root ganglion (Sugiura, Lee & Perl, 1986). Similarly to the lack of retrograde transport in the central nervous system, there is little evidence for the occurrence of retrograde transport of the leucoagglutinin in the peripheral nervous system. However, there are recent indications that the leucoagglutinin may be transported from the rat muscle motor neurons in a retrograde direction at about 8 mm/day which is comparable with that of the anterograde transport of the leucoagglutinin in the central nervous system. The primary afferent terminals are not labelled and dorsal rhizotomy does not reduce the extent of motor neuron labelling. Thus, transganglionic, transsynaptic PHA-L_4 transport by dorsal root ganglion cells does not occur and, therefore, this transport cannot be responsible for the labelling of the motor neurons. Hence, the leucoagglutinin can no longer be regarded as a specific anterograde transport tracer (Lee, McFarland & Wolpaw, 1988).

Suicide transport of toxic lectins. It appears that cytotoxic lectins are also taken up and transported within the axons. Lectins, such as ricin, abrin or modeccin, when injected into the distal vagus or hypoglossal nerves, are retrogradely transported along neuronal processes, where they inactivate ribosomes and cause death of neurons (Wiley, Blessing & Reis, 1982). For this reason, the retrograde transport of cytotoxic lectins is known as suicide transport.

Motor neurons can be degenerated experimentally by retrograde axoplasmic transport of ricin injected into the rat sciatic nerve and the intraaxonal transport of ricin to the motor neuronal soma and to dorsal root ganglia of L_{4-6} can be traced by immunohistochemical methods. The ganglion cells and motor neurons in sciatic nerve efferents of the lumbar spinal cord degenerate within a few days, while the small internuncial neurons and glia remain intact (Yamamoto *et al.*, 1985).

The cervical vagus retrogradely transports injected ricin to the cell body of the dorsal motor nucleus large neurons which are selectively destroyed by the toxin. As a result, a massive infiltration of mononuclear cells and macrophages occurs, similar to the situation in neural injury. The demonstration of numerous degenerating axon terminals and fibres in the dorsal motor nucleus indicates that ricin kills the vagal sensory neurons (Ling & Leong, 1987).

Injection of ricin into the rat sciatic nerve causes a massive degeneration of cells in the dorsal root ganglia which contain the sensory cells-of-origin of the sciatic nerve. With about 1–2 μg ricin injected, the average proportion of L_5 dorsal root ganglion neurons destroyed is similar to the number of axons of this ganglion in the sciatic nerve. With one-tenth of this dose, 0.2 g ricin, the extent of cell death is reduced, although heavy fibre degeneration occurs in the sciatic nerve, proximal to the injection site. In contrast, no degeneration takes place in a nearby tributary nerve sharing the L_5 dorsal root ganglion with the sciatic nerve. Accordingly, the lethal effects of ricin may be selective and complete on target neurons when used at the appropriate dose (Paul & Devor, 1987).

Cytotoxic lectins similar to ricin, such as modeccin and volkensin, may be more effective suicide transport agents than ricin in the peripheral nervous system (Wiley & Stirpe, 1987). Moreover, they may also destroy motor neurons adjacent to those which take them up and transport them; a spread of the degenerative lesion in the central nervous system which does not occur with ricin. Abrin, modeccin or volkensin, injected unilaterally into the caudate nucleus of rats produce extensive necrosis at the caudate injection site. Moreover, modeccin and volkensin, but not abrin, destroy neurons in the ipsilateral substantia nigra and intralaminar thalamus. The loss of dopaminergic neurons from the ipsilateral substantia nigra caused by the effects of modeccin and volkensin has been confirmed by histofluorescence. Thus, modeccin and volkensin are both effective suicide transport agents within the rat central nervous system presumably due to their retrograde axonal transport (Wiley & Stirpe, 1988).

Effects of lectins on the olfactory system. In studies on the binding patterns of different lectins conjugated to horseradish peroxidase to the nervous system of juvenile *Xenopus borealis*, only the agglutinin from soyabean has been found to be reactive. The binding is very selective and restricted to the olfactory system, with the olfactory and vomeronasal epithelia and the olfactory and accessory olfactory nerve bulbs becoming labelled. Moreover, the ventral portions of the olfactory nerve and bulb are more intensely labelled than the dorsal portions, while the rest of the brain and spinal cord remains largely unlabelled. In addition, the lectin is confined to the cell surface of olfactory neurons bound to specific cell surface glycoconjugates, which may have either functional or developmental roles in the olfactory system of *Xenopus* (Key & Giorgi, 1986).

Receptor cells of the olfactory system of vertebrates probably have a fairly wide range of responses and it is thought that most of these cells can respond to several odorants. However, so far, the response spectrum of the cells of the olfactory system is not clear and even less is known about the actual receptor molecules. It has been suggested that a typical odorant may interact with several different receptor molecules and that the actual odour is recognized from the extent and the relative degrees of the stimulation of the receptors. As lectins react with receptors on cell membranes, it is not surprising that, after blockage with lectins, some of the receptors may become inactivated. For example, concanavalin A superfused over the olfactory mucosa *in vitro* can affect responses to some odours; with rat olfactory mucosa, concanavalin A reduces the sensitivity of response to about 60% of the 112 odorants tested. Thiols, carboxylic acids and hydrocarbons containing four to six carbon atoms are particularly affected by the administration of the lectin suggesting a weak structural specificity of the receptor for this kind of odorant. It is possible that concanavalin A may bind at, or close to, one or more of the odorant receptors (Shirley *et al.*, 1987*a*). Concanavalin A reduces the olfactory response to hydrophobic odorants by concanavalin A dependent on their concentration. It is suggested that this is compatible with the idea that concanavalin A may inactivate one or, possibly, several types of the olfactory receptors which normally bind odorants with dissociation constants of the order of about 100 nM. Very much higher concentrations of the hydrophylic odorants are needed to obtain a reduction of the response by concanavalin A (Shirley *et al.*, 1987*b*).

Miscellaneous interactions of lectins with cells and their underlying structural elements

The binding of plant lectins to simple carbohydrates or complex carbo-hydrate moieties of cell surface glycoconjugates provides a useful tool for the investigation of the relationship between receptor clustering, follow-ing the attachment of growth factors to cell membranes and the effects of the lateral movement of membrane receptors on the underlying cytoplas-mic structures. The importance of such an approach is highlighted by the findings that receptor clustering is an important early step in the mito-genic response of cells (King & Cuatrecasas, 1981; Schreiber *et al.*, 1983). It is clear that lectins can restrict receptor mobility (Yahara & Edelman, 1975; Henis & Elson, 1981; Schlessinger *et al.*, 1977) and it may be expected that lectins interfere with the mitogenic effects of peptide growth factors on cells, due not only to their competitive binding to cell surface receptors, but also as a result of their interference with receptor clustering. These effects are not to be mistaken for the already well-known insulin- or somatomedin C-like or mitogenic effects of lectins on various cell lines which are due to the occupation of peptide growth factor receptor sites by lectins. It is known, for example, that, although both wheatgerm agglutinin and concanavalin A bind to the epidermal growth factor receptor on the membranes of fibroblasts (Carpenter & Cohen, 1977; Kaplowitz, 1985), these lectins do not mimic all the effects of somatomedin C, insulin or epidermal growth factors on the cells and, most importantly, they do not stimulate DNA synthesis. However, as judged from their stimulatory effects on such early events in mitogenic stimulation as the increased uptake of aminoisobutyric acid (a non-metabolized amino acid) or amino acids, the binding of wheatgerm agglutinin or concanavalin A appears to cause receptor cross-linking in the membranes of human fibroblasts. Accordingly, by the treatment of these cells with wheatgerm agglutinin or concanavalin A, the early events of amino acid uptake can be dissociated from the late phases of mitogene-sis, i.e. the stimulation of DNA synthesis (Kaplowitz, 1985), suggesting that amino acid uptake in itself is not sufficient to allow the cells to enter the *S* phase of the cell cycle.

However, despite the lack of full mitogenic effects of their own, wheatgerm agglutinin or concanavalin A can still inhibit not just the early but also some of the late events of mitogenic stimulation of various peptide growth factors in human fibroblasts. Even more significantly, wheatgerm agglutinin or concanavalin A inhibits both the uptake of aminoisobutyric acid and DNA synthesis induced by growth factors in a non-competitive way. Thus, this inhibition cannot be reversed by as much as a hundredfold excess of insulin. Moreover, in relation to the application of the peptide growth factors, delaying the addition of concanavalin A does not impair the ability of the lectin to inhibit the synthesis of DNA induced by, for example, epidermal growth factor. The divalent succinyl-concanavalin A not only fails to block the mitogenic effects of the epidermal growth factor but it actually enhances it, despite its potent inhibitory effect on the binding of epidermal growth factor. This suggests that inhibition by lectins of the mitogenic reaction is, indeed, independent of their interference with the attachment step in the induction of the mitogenic reaction by the growth factors (Kaplowitz, 1985).

It appears that the reason for the unresponsiveness of human fibroblasts to mitogenic stimulation by peptide growth factors in the presence of lectins cannot be ascribed to competitive binding of the lectins to the growth factor receptors. Thus, other events, such as the inhibition by lectins of the lateral mobility of membrane receptors, may be relevant. This is supported by the inability of the divalent succinyl-concanavalin A to cross-link mobile concanavalin A-receptors effectively (Gunther *et al.*, 1973). Consequently, the divalent lectin is unable to inhibit the growth factor-induced DNA synthesis in fibroblasts. Thus, receptor aggregation may indeed be a crucially important step in the late phases of the mitogenic stimulation by peptide growth factors.

The aggregated lectin–receptor complexes may also disturb the structure of cytoplasmic microtubules and microfilaments attached to the membrane receptors. The cytoplasmic distortion thus caused may effectively inhibit the cell's responsiveness to the mitogenic stimulus of the growth factors. Indeed, some experimental evidence appears to support such ideas. For example, microtubule-disruptive agents can inhibit receptor clustering induced by lectins in both lymphocytes and 3T3 cells (Yahara & Edelman, 1975; Schlessinger *et al.*, 1972). Moreover, lectins can also affect the interaction between membrane glycoproteins and the cytoskeleton in, for example, platelets (Painter & Ginsberg, 1982) or some cells (Vale & Shooter, 1983).

There have been several attempts to rationalize these observations and the curious finding that, in fibroblasts, which have been exposed to insulin or somatomedin C for 1 hour before wheatgerm agglutinin is introduced into the reaction mixture, the uptake of aminoisobutyric acid is inhibited to nearly the same extent as when the lectin is added to the cells prior to the growth factors (Kaplowitz, 1985). Both ligand-induced receptor clustering into coated pits and the following endocytosis into receptosomes, are relatively fast reactions (Pastan & Willingham, 1981). Consequently, wheatgerm agglutinin, which is administered after the initial attachment of the growth factors, can disrupt only those ligand–receptor complexes which have not yet been internalized. Despite this, the lectin is fully effective in inhibiting the early events of mitogenesis, such as the uptake of aminoisobutyric acid. Accordingly, the mitogenic effects of the growth factors must be due to those ligand–receptor complexes which remained on the cell surface and not to those which have been endocytosed via receptosomes. Alternatively, by sending their own signals into the interior of the cell, it is also possible that the lectin–receptor clusters on the cell surface may negate the mitogenic signals of the peptide growth factors, irrespective of whether or not they have been endocytosed.

Indeed, some of the evidence seems to point to the second possibility as the most likely way lectins interfere with the growth factor-induced mitogenic response. For example, wheatgerm agglutinin at concentrations high enough to almost completely suppress epidermal growth factor-induced DNA synthesis in fibroblasts has practically no effects on the endocytosis of the growth factor by the cells. Even concanavalin A has only a relatively minor effect on the internalization of the epidermal growth factor by fibroblasts. Also, wheatgerm agglutinin added to fibroblasts as late as 16 to 20 hours after the growth factor, can still inhibit the synthesis of DNA by the cells. Thus, the blockage by the lectin of the mitogenic pathway must occur late in the G_1 phase of the cell cycle. From thymidine-incorporation experiments, in which wheatgerm agglutinin is co-applied with anti-epidermal growth factor antibodies, the time-scale of the interference by the lectin has been further narrowed to about 6–8 hours after the attachment of the growth factor, close to the G_1/S boundary of the cell cycle. Moreover, this time point for the inhibition of DNA synthesis by the lectin is also consistent with the observations that wheatgerm agglutinin inhibits the mitogenicity of non-binding mitogens, such as colchicine or sodium orthovanadate (Kaplowitz & Haar, 1988).

Wheatgerm agglutinin bound to surface receptors of the fibroblasts remains associated with the membrane for a long time without significant endocytosis in contrast to the fast ligand–receptor internalization occurring after peptide growth factor binding. This persistent binding of wheatgerm agglutinin to the cell surface is both necessary and sufficient for the inhibition of the mitogenic effects of the growth factors. In accordance with this, by washing the lectin-treated cells with N-acetyl-D-glucosamine, the membrane-bound wheatgerm agglutinin is rapidly removed. This, in turn, restores the response of the cell to the stimulation of DNA synthesis by epidermal growth factor (Kaplowitz & Haar, 1988).

Despite the great advances in our understanding of the reaction mechanism of the inhibition by lectins of the mitogenic actions of peptide growth factors on human fibroblasts, the precise details of lectin action is still not clear. Moreover, the complex effects of lectins, such as wheatgerm agglutinin or concanavalin A, on the mitogenic stimulus indicate the existence of several mitogenic pathways through which the various growth factors can stimulate each successive step in the mitogenic transformation of cells. It is particularly significant that membrane-bound wheatgerm agglutinin can affect and interfere with DNA synthesis induced by the internalized growth factor-receptor and that this is fully reversible. As a result of the persistence of the lectin-induced receptor clustering and the ensuing changes in the cytoplasmic scaffolding, both wheatgerm agglutinin and concanavalin A induce relatively long-lasting changes in cell shape and affect cell spreading. These changes in the internal structure of the cell may, in themselves, be responsible for the reduced mitogenic responsiveness. Alternatively, shifts and alterations in the relative chemical composition of the membrane can occur due to receptor clustering. Consequently, the modified membrane can no longer send the right signal to the interior of the cell for the stimulation of DNA synthesis.

The importance of changes in cell shape and shifts in membrane and cytoplasmic structural components is also highlighted by the different responses of Swiss 3T3 fibroblasts or L_6 myoblasts to physiological concentrations (ng/ml) of insulin or *Phaseolus vulgaris* lectin, PHA-L_4. Thus, equimolar concentrations of both insulin and PHA stimulate the synthesis of cellular proteins in 3T3 fibroblasts and the two stimuli appear to be additive (Table 6.10). In contrast, while insulin is a powerful stimulant of protein synthesis for L_6 cells, PHA on its own does not increase protein synthesis above non-stimulated control values in these cells. However, when both factors are applied together, the combined rates of protein synthesis exceed significantly those obtained with insulin alone, particularly at late stages of stimulation (Table 6.10).

It is possible that the different effectiveness of these two signals may find an explanation in the morphological changes induced by fluorescein-labelled insulin or fluorescein-labelled PHA (Figs 6.23 & 6.24) in non-fixed preparations of these two cells. The surface membranes of these cells react practically instantaneously with both stimulants. Extensive clustering and patching of surface receptors occurs with either insulin or PHA within 10–30 min of the reaction. However, significant amounts of the fluorescein-labelled insulin are endocytosed by both cells. In fact, the

Table 6.10. *Stimulation of protein synthesis by insulin or PHA-L$_4$ lectin in* (a) *Swiss 3T3 fibroblasts or* (b) *L$_6$ myoblasts.*

Time (h)	Insulin	PHA-L$_4$	Insulin + PHA-L$_4$
(a)			
0–1	18[a]	2	14
2–3	24[a]	19[a]	33[a]
7–8	35[a]	13[a]	43[a]
(b)			
2–3	45[a]	0	55[a]
4–5	55[a]	1	64[a]
7–8	55[a]	0	79[a]

Note: The results are given as the difference between percentage incorporation of [^{3}H]-phenylalanine into stimulated (2 ng/ml insulin or lectin) or control cells; [a] significantly different from controls, $p < 0.05$.

endocytosed insulin effectively stains the nuclei of both 3T3 and L$_6$ cells and this direct interaction of insulin with the nucleus or nuclear membranes probably accounts for its long-lasting effects on protein synthesis. Such a direct stimulation of protein synthesis through nuclear attachment of growth factors may be particularly important in the second of the two separate phases of increased protein synthesis observed in both cells (Bardocz & Hesketh, 1989). The effects of PHA, however, are different (Figs 6.23 and 6.24). Although the lectin stimulates protein synthesis in 3T3 cells, it is less effective than insulin. Moreover, the lectin remains attached to the membrane of these cells and it is apparently not endocytosed.

In L$_6$ myoblasts, PHA does not stimulate protein synthesis at all nor is it endocytosed to a significant extent by L$_6$ cells either. However, the surface-bound PHA has striking effects on the membrane structure and cell shape. Moreover, the deformation of the cell shape caused by the lectin becomes accentuated with time (Fig. 6.23) and this may indeed have a contributory effect to the significantly higher stimulation of protein synthesis achieved by the combined application of the two stimulants on L$_6$ cells. This is further supported by the finding that, when PHA is added to the cells before insulin, the nuclear staining occurs much faster than that observed with insulin alone. In this respect, the effect of unlabelled PHA on these cells is similar to that of fixation carried out with organic solvents before exposure of the cells to fluorescein-labelled insulin. In fact, with such pre-fixation, the cell membrane becomes

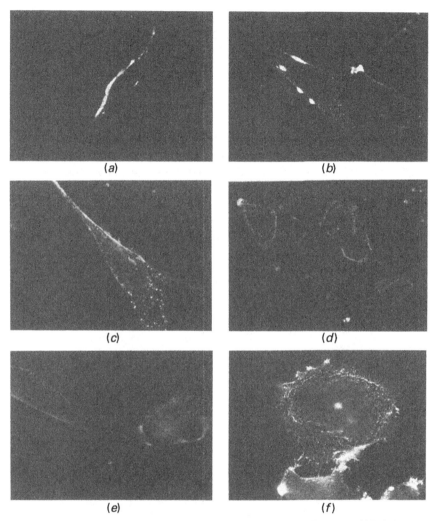

Fig. 6.23. Binding and endocytosis of PHA by Swiss 3T3 fibroblasts ((*a*)–(*c*)) and L_6 myoblasts ((*d*)–(*f*)). Cells were incubated with FITC-labelled PHA (2 ng/ml) for: 1 min ((*a*) and (*d*)), 10 min ((*b*) and (*e*)) or for 30 min ((*c*) and (*f*)), washed with distilled water and fixed with 4% paraformaldehyde in saline for 10 min. Scale bar represents 20 μm.

almost totally and instantaneously permeable to subsequently applied insulin.

The changes observed in these studies are, therefore, different from those found in rat hepatoma cells stimulated by ^{125}I-labelled insulin. Thus, treatment with concanavalin A after the insulin stimulus, increases the binding of the insulin to the cells at 23 °C but not at low temperatures.

However, these cells endocytose concanavalin A and insulin together (Sorimachi, Okayasu & Yasumura, 1987). Moreover, in contrast to the increased membrane permeability found with PHA pre-treatment described previously (p. 194), when concanavalin A is applied before insulin, it reduces insulin-binding not only by steric hindrance but also because the concanavalin A receptors that are internalized are identical with the insulin-binding sites. It is, therefore, clear that lectins and growth factors applied either separately or together may have different effects on membrane components or underlying structural elements in different cell lines. This is probably largely determined by differences in

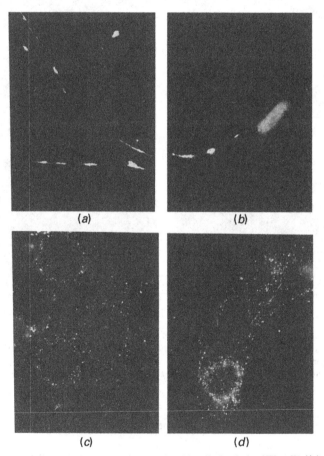

Fig. 6.24. Binding and endocytosis of insulin by Swiss 3T3 cells ((a)–(c)) or L$_6$ myoblasts (d). Cells were incubated with FITC-labelled insulin (2 ng/ml) for 1 min (a), 10 min (b) or for 30 min ((c) and (d)), then washed with distilled water and fixed with 4% paraformaldehyde in saline for 10 min. Scale bar represents 20 μm.

the receptors of such ligands and their numbers or availabilities. The response of the cell is mainly dependent on how these receptors are anchored to the cytoskeletal elements and the possibility of freedom of their lateral movements within the plane of the membrane. Thus, it is likely that these largely unknown effects of the different growth factors or lectins on the cell membranes and underlying cytoplasmic structures will largely pre-determine the overall response of the cell to the stimulus.

Similarly, the relationship between the changes in cell shape and cell spreading induced by growth factors and the more thoroughly studied other mitogenic cellular responses, such as the increased cellular protein synthesis, is largely unknown. Although they may be linked, it is also possible that the two phenomena are just unrelated changes of the initial binding.

The effects of lectins on cell binding and spreading on a laminin substrate can be measured directly. Exposure of laminin to wheatgerm agglutinin inhibits the binding of mouse B16 F1 melanoma cells to this prominent component of the basement membrane (Dean, Chandrasekaran & Tanzer, 1988), whereas exposure of this cell to concanavalin A does not affect the attachment but inhibits cell spreading on laminin. This effect on the cell–laminin interaction is clearly due to the attachment of the lectins to the carbohydrate side-chains of laminin since the inhibition of both binding and spreading can be prevented by the addition of sugars which are specific for the lectins. As they are already bound to laminin before the addition of the cells, the main effects of the lectins are probably on the laminin and are not due to their attachment to or the consequent deformation of the cells. However, the possibility of such a direct lectin effect on the cells cannot be excluded completely in these experiments (Dean *et al.*, 1988). This is even more important in view of previous observations which have shown that the attachment of tumour cells to the basal lamina is a pre-requisite for metastasis. This attachment to laminin or fibronectin is through specific receptors on the cells themselves (Lang *et al.*, 1987). Thus, it is expected that lectins may interfere with the interaction between cells and laminin not just through their binding to laminin but also by their attachment to those glycoconjugates of cell surfaces which are involved in the mediation of cell-binding to laminin.

There is increasing evidence to show that membrane receptors of lectins are associated with the underlying actin filaments in the cytoplasm. For example, the existence of filamentous structures consisting of fine actin filaments, microtubules and endoplasmic reticulum has been revealed by stereo electron microscopy of specific antibody-stained, saponin extracted and partially opened, well-spread monolayers of rat 3Y1 cells. The actin-containing fine filaments are linked to microtubules and the endoplasmic reticulum just beneath the plasma membrane. Some of these are associated with the cytoplasmic surface of the plasma membrane either by their ends or by their lateral surfaces and the presence of membrane receptors for concanavalin A can be demonstrated by binding colloidal gold-labelled concanavalin A to the surface of these saponin-extracted fixed 3Y1 cells. Moreover, virtually all gold particles bind externally to the same membrane sites to which the intracellular actin filaments are attached internally. Thus, the distribution of lectin receptors in the membranes and their involvement in lectin-induced patching and capping is obviously regulated by the underlying meshwork of actin filaments (Katsumoto & Kurimura, 1988).

The existence of an intimate relationship between surface receptors and the underlying actin filaments has also been confirmed. Thus, the exposure of Caco-2 cells to soyabean agglutinin causes the depolymerization of F-actin within the cell. The appearance of G-actin is dependent on the amounts of lectin present. Significantly, in the presence of the carbohydrate, for which soyabean agglutinin is specific, *N*-acetyl-D-galactosamine, depolymerization does not occur (Draaijer *et al.*, 1989).

Changes in the morphology of the cytoskeletal structure also lead to shortened and disrupted intestinal microvilli in the lectin-exposed rats (King *et al.*, 1982; Lorenzsonn & Olsen, 1982; Sjölander *et al.*, 1986). Indeed, as one of the most conspicuous effects of soyabean agglutinin on Caco-2 cells is a shortening of the microvilli (Draaijer *et al.*, 1989), the *in vitro* model studies with these cells may open the way for exploring the *in vivo* morphological and functional effects of lectins on small intestinal enterocytes.

To examine the role of the cytoskeleton in concanavalin-A-induced receptor patching and capping, the soluble portion of the cell needs to be removed first by a Triton X-100-solubilization procedure, which leaves the intact cytoskeletal structure behind as an insoluble residue (Gilbert & Fulton, 1985). This technique has been used widely. For example, the clustering of laminin by lectins in fibrosarcoma cells induces the binding of laminin to the cytoskeleton via the transmembrane protein, connectin (Cody & Wicha, 1986). In a similar way, the solubilization procedure with Triton X-100 identifies specific glycoproteins which are transferred to the Triton-insoluble cytoskeletal fraction after the initial attachment of concanavalin A to membrane receptors. Some glycoproteins are always associated with the cytoskeleton, regardless of whether the cells have been exposed to concanavalin A or not. However, after concanavalin A-induced capping of cells of *Dictyostelium discoideum*, two additional classes of cell membrane glycoproteins can be identified in the Triton X-100-insoluble fraction. One of these, containing class 2 proteins, associates immediately with the cytoskeleton, while another, a so-called class 3 protein, partitions into the detergent-insoluble fraction only after extensive cell surface glycoprotein cross-linking (Patton, Dhanak & Jacobson, 1969). The rapid binding of class 2 proteins to the cytoskeleton after the initial attachment of concanavalin A is accompanied by dramatic changes in cell shape, the amoeboid changes into a more rounded form with the cessation of active extension and retraction of filopodia. It is likely that class 2 membrane proteins may play an important part in the initiation of cellular responses to the lectin. In contrast, the slow-reacting class 3 membrane proteins are probably involved only in late cellular reactions to the lectin. However, the precise mechanism of the various steps in the reactions of membrane proteins with cytoskeletal components after the binding of lectins and the reactions which lead to patching and capping of membrane components are not yet clear.

Cytotoxins – magic bullets – immunotoxins

Aub and his associates (1963) have shown that the binding of wheatgerm agglutinin to transformed cells is more extensive than that of the same cells before transformation. Ever since then, the use of various lectins as tools for the selective delivery of toxins to tumour cells without killing the normal, healthy cells has been explored. Wheatgerm agglutinin is not toxic enough to kill most cells, but the extremely poisonous ricin-type lectin toxins and the ribosome-inactivating polypeptides which occur widely in Nature can be used for this purpose after their successful delivery into the cell cytosol. As probably a single molecule of the toxic A-chain of ricin is sufficient to cause cell death, the practical problem is not that of the killing but rather that of the selective delivery.

Although intact ricin reacts with, and kills, ascites tumour cells, the specificity of the delivery agent, the galactose-specific B-chain, is too broad to be of practical value for selective killing (Olsnes & Pihl, 1977) or for clinical use (Fodstad *et al.*, 1984). The selectivity of the delivery can be increased by removing the B-chain lectin from the intact A–B molecule and chemically linking the toxic A-chain to monoclonal antibodies specific for the particular type of cell. The resulting molecules, called immunotoxins, are much more specific than the intact ricin; they can recognize and bind to cells carrying the surface antigen against which the monoclonal antibody has been raised. Although immunotoxins made by the removal of the lectin B-chain are, in theory, easier to target and more specific than the native ricin, they are not lectins. Therefore, they are outside the scope of this book.

The potency of immunotoxins is low and unpredictable if the lectin B-chain is not retained. In contrast, immunotoxins made by coupling the whole ricin molecule to monoclonal antibodies are much more potent. The presence of the lectin B-chain therefore appears to be necessary for the translocation of the toxic A-chain into the target cell cytosol.

The initial attachment of the immunotoxin to its membrane receptors needs to be explored first in order to understand the precise mechanism of toxin delivery to its target. In addition, the intricate successive steps of the intracellular movement of the immunotoxin molecule and its chemical and enzymic reactions with components of the various subcellular compartments, also need to be understood. Obviously, as non-toxic lectins are also endocytosed by a mechanism similar to that of the ricin molecule, some of this understanding may come from studies of mapping the ways of intracellular traffic of the less toxic lectins, such as wheatgerm agglutinin.

Thus, horseradish peroxidase-labelled wheatgerm agglutinin administered intravenously was identified in endocytic vesicles, spherical endosomes, tubules, dense and multivesicular bodies, the *trans*-most saccule of the Golgi apparatus and dense-core secretory granules attached to the *trans*-Golgi saccule in anterior pituitary cells or in some neurons. On the other hand, no label was found in the endoplasmic reticulum and the *cis* or intermediate Golgi stacks (Balin & Broadwell, 1987). However, these results need more supporting evidence. It is possible that the indicated endocytic-exocytic pathways in pituitary cells, i.e. the eventual conveyence of the internalized cell membrane receptor–horseradish peroxidase–wheatgerm agglutinin complex to the *trans*-most Golgi stacks, followed by their packaging in dense-core secretory granules for export and exocytosis, is the result of the presence of the horseradish peroxidase label and not the lectin molecule.

Tritium (^{3}H), on the other hand, is one of the least perturbing labels which can be used for following the steps of endocytosed lectins and locating them in the various subcellular compartments. [^{3}H]-labelled wheatgerm agglutinin is particularly useful as it is unlikely that its intracellular movement would be subject to the same handicap as that due to its derivatization by large marker molecules (van Deurs *et al.*, 1986). Quantitative ultrastructural and autoradiographic studies with [^{3}H]-wheatgerm agglutinin have confirmed that the Golgi apparatus does indeed participate in the intracellular transport of the lectin in cultured murine neuroblastoma cells (Mezitis, Stieber & Gonatas, 1987). The intracellular route of the lectin has been followed by incubating the cells with the labelled lectin for one hour. Subsequently, the cells have been washed with the complete medium without the ligand for various time periods and the grain density in each intracellular compartment has been measured. For the first two hours, most of the label is associated with tubules, smooth vesicles, Golgi-associated vesicles and cisterns, while lysosomes are only poorly labelled. After two hours, most of the lectin is located in the lysosomes although there is still some label in the Golgi-associated vesicles, indicating that the Golgi complex (cisterns and associated vesicles) is an early or intermediate step in the endocytosis of [^{3}H]-wheatgerm agglutinin *en route* to the lysosomes. Alternatively, the results are also consistent with the idea that the Golgi complex constitutes a quantitatively significant part of a separate, non-lysosomal pathway in the endocytosis of wheatgerm agglutinin by neuroblastoma cells (Mezitis *et al.*, 1987).

The involvement of the Golgi complex in the intracellular routing of wheatgerm agglutinin has been confirmed not just in higher organisms but also in ciliated protozoa, such as *Paramecium*, by the use of cryo-sectioning combined with pulse-chase studies and wheatgerm agglutinin–gold labelling (Allen, Schroeder & Fok, 1989). Binding of wheatgerm agglutinin to the *trans* face of the Golgi stacks is followed by the binding of the lectin to the entire phagosome–lysosome system of the protozoa. This is consistent with ideas concerning the role of the Golgi apparatus in the terminal steps of glycosylation of membrane glycoproteins and acid hydrolases destined for the digestive system of the *Paramecium*.

The existence of a non-lysosomal pathway of endocytosis, in which the Golgi apparatus is involved, may be crucially important for the delivery of the ricin part of the immunotoxins to the cytosol and cell death. As we have seen it before (p. 179), a low pH-induced conformational change and dissociation appears to be necessary for the translocation through the endosomic membrane into the cytosol of the endocytosed diphtheria A–B type toxins. In contrast, intact ricin or ricin-containing immunotoxins are probably transported into the cytosol via the Golgi apparatus by a different mechanism. Thus, for its delivery into the cytosol, ricin may utilize the link existing between the endosomal system-based inward membrane traffic and the Golgi-dependent outward (exocytic) membrane transport (van Deurs *et al.*, 1986, 1987, 1988). It is estimated that about 5% of the total amounts of the ricin internalized by the cells reaches the Golgi complex within 60 minutes of the initial exposure of BHK-21 cells to the lectin. Most of this (70–80%) is present in the *trans*-Golgi network. Thus, a small but distinct fraction of the internalized ricin avoids the usual major lysosomal route for the disposal of endocytosed molecules. This part of the endocytosed ricin is not degraded within the cell but it may be translocated to the cytosol from the *trans*-Golgi network (van Deurs *et al.*, 1988). Accordingly, both the relatively non-toxic wheatgerm agglutinin and the extremely toxic ricin appear to be processed through a similar endocytic pathway of which the main functional component used for exocytosis is the *trans*-Golgi saccules. The main difference in behaviour between the two lectins is in the final step, after their delivery to the cytosol. Although the further fate of wheatgerm agglutinin in the cytosol is unknown, the toxic A-chain of ricin halts the synthesis of proteins and causes cell death by inactivating the 60S ribosomal subunits.

From these studies of the endocytic pathways of internalized toxins, it appears that one of the main requirements for the success of the delivery of the toxin to the cytosol is the presence of the B-chain of the lectin. This probably explains why immunotoxins made by linking the entire toxin molecule of ricin to the monoclonal antibody are at least as effective as toxins for the cells as the native ricin. Such immunotoxins, unfortunately, suffer lack of strict specificity which is a major disadvantage. The immunoreagent binds to all cells bearing galactose-containing surface glycoconjugates due to the presence of the galactose-specific B-chain. However, these types of immunotoxins are still useful for *in vitro* clinical applications. For example, T-cells or tumour cells may be removed from the bone marrow before allogeneic or autologous bone marrow transplantation (Pastan, Willingham & FitzGerald, 1986). This can be done by blocking the galactose binding sites on the B-chain with lactose (Youle & Neville, 1980) or galactose (Thorpe *et al.*, 1981). Clearly, because of the rapid removal of galactose or lactose by body metabolism, this strategy of blocking the reactivity of the B-chain cannot be achieved *in vivo*. In a more sophisticated approach, the ricin B-chain binding site can also be blocked sterically by the antibody used for the targeting. Thus, an anti-Thy 1.1 conjugate, in which the lectin function of ricin has been blocked by the antibody, is four orders of magnitude more cytotoxic to a cell line expressing Thy 1.1 than to cell lines expressing Thy 1.2 antigens (Thorpe *et al.*, 1984). All AKR-A cells carrying the Thy 1.1 antigen from the peritoneum of mice can be totally and selectively removed by a single injection of this immunotoxin (Thorpe, 1985).

However, the most obvious way to minimize the non-specific and general reactions of the immunotoxins with cells is to couple the toxic A-chain to monoclonal antibodies, instead of using the whole ricin molecule for the coupling. By expressing the cloned A-chain gene in *E. coli*, the preparation of completely pure, non-glycosylated A-chain is feasible. Such immunotoxins can now be prepared without any risk of contamination with the lectin B-chain.

The effectiveness of A-chain immunotoxins is rather unpredictable. One of the reasons for this poor performance is that the glycosylated A-chain is rapidly cleared by liver cells which contain endogenous lectins specific for mannose and fucose residues present on the A-subunit. Although it is not very clear how the presence of the B-chain facilitates this process and by what mechanism, it appears that the A-chain-containing immunotoxins are not translocated into the cytosol with the same high efficiency as those which also contain the lectin B-subunit. One such reason may be that the presence of the B-chain stabilizes the immunotoxin against proteolytic breakdown, since during the intracellular passage the survival of immunotoxins made with intact ricin is more extensive than that of the less stable A-chain immunotoxins. Indeed, one of the successful strategies used for increasing the stability of immunotoxins is to use agents which increase the pH of the phagolysosomes (Casselas *et al.*, 1982) or alter the intracellular pathways of the immunotoxin, such as ammonium chloride, chloroquine and other lysosomotropic agents. Accordingly, both the carboxylic ionophore, monensin (Casselas *et al.*, 1984) and the calcium channel blocker verapamil (Pirker *et al.*, 1988) enhance the activity of the ricin A-chain conjugate-containing immunotoxins.

The B-chain may also be necessary for forming channels or tunnels in the membrane through which the A-chain can pass into the cytosol (Blaustein *et al.*, 1987). Rather interestingly, when the two reagents are applied together, even the presence of a separate B-chain appears to potentiate the biological activity of the A-chain conjugate immunotoxin (Youle & Neville, 1982). The presence of a separate B-chain conjugate has the same effect and enhances the A-chain immunotoxin cytotoxicity *in vitro* (Vitetta, Cushley & Uhr, 1983). Similarly, the B-chain conjugated with an antibody which is directed against the antibody part of the A-chain immunotoxin is also effective in potentiating cytotoxicity (Vitetta, Fulton & Uhr, 1984).

Immunotoxins have already been used in clinical practice despite these very real and, at present, not fully solved problems. Applications of a number of immunotoxins have been described, befitting their great potential importance and usefulness (Vitetta & Uhr, 1984; Lord *et al.*, 1985; Lord, 1987; Drobniewski, 1989). Although a detailed description of the clinical use of immunotoxins is obviously outside the scope of this book, some of the problems encountered when immunotoxins are used in *in vivo* therapy have a close relevance to the use of lectins generally and are described here briefly.

When intact ricin-containing immunotoxins are used *in vivo*, the stability of the covalent bond linking the toxin to the antibody is of great importance. For the prevention of indiscriminate toxicity, the linkage must be very stable during delivery to the cell, but should be relatively easy to break once the toxin is inside the cell. However, the stability of the immunotoxin is important not just when it is made of intact ricin, but also when the conjugate is based on the A-chain. Thus, the monoclonal antibody liberated from an injected unstable immunotoxin may reduce the effectiveness of the immunotoxin by competing with the intact agent. In addition, as the conjugate is more rapidly cleared by the liver than the free monoclonal antibody, this competition is likely to increase with time. Fortunately, there are now several reasonable reliable coupling agents available for producing stable conjugates in practice (Thorpe *et al.*, 1987, 1988) and it is likely that even more useful chemical reactions will be found in future.

Another problem which may arise is that the tumour cells may naturally shed some of their surface antigens against which the monoclonal antibodies have originally been produced and this can reduce the efficacy of the specific killing of target cells by the immunotoxins. The surface density of suitable antigens may also be reduced substantially by a rapid internalization of the receptors. These problems can be avoided by using monoclonal antibodies raised against the more stable surface antigens of target cells.

As both the antibody and the toxin are immunogenic proteins, antibodies may be produced against one or the other (or both) components of the conjugate by the immune system of the recipient during extended usage in cancer therapy (Spitler *et al.*, 1987). The neutralizing effect of the humoral antibody response (IgG) or the formation of allergenic, IgE-type, antibodies may seriously reduce the effectiveness of the immunotoxin in the cancer patient. Some of the problems due to the immunogenicity of the antibody part of the conjugate, except the anti-idiotype response, may be avoided by the use of human monoclonal antibodies. Other immunologically unrelated toxins, such as the ribosome-inactivating proteins, may also be used in successive regimes of therapy instead of the ricin A-chain. Under these conditions, the efficiency of immunotoxin therapy may be maintained at a high level for extended periods of treatment. It may be generally advantageous to use several distinct immunoconjugates owing to the observed resistance of some tumour cells to respond well to singular immunotoxins. It is of particular advantage if, in the immunotoxins, both antibodies are directed to several antigens on the tumour cell and the toxins are also different.

The discovery of magic bullet immunotoxins and their exploration for clinical use is one of the recent major advances in medical science and clearly owes a heavy debt to lectinology. Although the delivery to target cells based on the lectin part of A–B toxic lectins is not specific enough for practical applications and it has now been superseded by the more precise targeting achieved with monoclonal antibodies, the presence of the lectin B-subunit still confers superior potency of cell-killing to immunotoxins which contain it in a way which is not understood. It is possible that one of the ways ahead to make more effective immunotoxins based on the whole ricin molecule will be to reduce most of the unwanted and non-specific binding of the lectin B-chain in a reliable way. This may be achieved through site-directed mutagenesis of the gene coding for the lectin part. Hopefully, while the non-specific binding of the immunotoxins to cells through such transformed B-chains can be reduced, the superior intra-cellular routing of the immunotoxin due to B-chain lectin function will still be retained.

References

Agrawal, B.B.L. & Goldstein, I.J. (1967). Specific binding of concanavalin A to cross-linked dextran gels. *The Biochemical Journal*, **96**, 23–5C.

—(1968). Protein carbohydrate interaction. XIV. The role of bivalent cations in concanavalin A–polysaccharide interaction. *Canadian Journal of Biochemistry*, **46**, 1147–50.

Alecio, M.R. & Rando, R.R. (1982). Threshold effects on the lectin-mediated aggregation of liposomes: Influence of the diameter of the liposomes. *Journal of Membrane Biology*, **67**, 137–41.

Aleksidze, G.Y., Korolev, N.P., Semenov, I.L. & Vyskrebentseva, E. I. (1983). Isolation of lectins and their possible receptors from sugar beet roots. *Fiziologiya Rastenii*, **30**, 1069–76.

Aleksidze, G.Y. & Vyskrebentseva, E.I. (1986). Subcellular localization of lectins in sugar-beet root tissues of various ages. *Fiziologiya Rastenii*, **33**, 213–20.

Allen, A.K. (1979). A lectin from the exudate of the fruit of the vegetable marrow (*Cucurbita pepo*) that has a specificity for β-1.4-linked N-acetylglucosamine oligosaccharides. *The Biochemical Journal*, **183**, 133–7.

Allen, A.K. & Neuberger, A. (1973). The purification and properties of the lectin from potato tubers, a hydroxproline-containing glycoprotein. *The Biochemical Journal*, **135**, 307–14.

—(1975). A simple method for the preparation of an affinity absorbent for soybean agglutinin using galactosamine and CH-Sepharose. *FEBS Letters*, **50**, 362–4.

Allen, A.K., Neuberger, A. & Sharon, N. (1973). The purification, composition and specificity of wheat germ agglutinin. *The Biochemical Journal*, **131**, 155–62.

Allen, A.K., Desai, N.N., Neuberger, A. & Creeth, J.M. (1978). Properties of potato lectin and the nature of its glycoprotein linkages. *The Biochemical Journal*, **171**, 665–74.

Allen, R.D., Schroeder, C.C. & Fok, A.K. (1989). Intracellular binding of wheat germ agglutinin by Golgi complexes, phagosomes and lysosomes of *Paramecium multimicronucleatum*. *The Journal of Histochemistry and Cytochemistry*, **37**, 195–202.

Alvarez, J.R. & Torres-Pinedo, R. (1982). Interactions of soyabean lectin soyasaponins and glycinin with rabbit jejunal mucosa *in vitro*. *Pediatric Research*, **16**, 728–31.

Andersen, M.M. & Ebbesen, K. (1986). Screening for lectins in common foods by line-dive immunoelectrophoresis and by haemadsorption lectin test. In

Lectins, Biology, Biochemistry, Clinical Biochemistry (Bog-Hansen, T.C. & van Driessche, E. eds), vol. 5, pp. 95–108. Walter de Gruyter, Berlin and New York.

Anderson, P.J., Bardocz, S., Campos, R. & Brown, D.L. (1985). The effect of polyamines on tubulin assembly. *Biochemical and Biophysical Research Communications*, **132**, 147–54.

Andrews, A.T. & Jayne-Williams, D.J. (1974). The identification of a phytohaemagglutinin in raw navy beans (*Phaseolus vulgaris* L.) toxic for the Japanese quail (*Coturnix coturnix japonica*). *The British Journal of Nutrition*, **32**, 181–8.

Angelisova, P. & Haskovec, C. (1978). Isolation and chemical characterisation of a highly purified phytomitogen from *Phaseolus coccineus* seeds. *European Journal of Biochemistry*, **83**, 163–8.

Araki, T., Yoshioka, Y. & Funatsu, G. (1986). The complete amino acid sequence of the B-chain of the *Ricinus communis* agglutinin isolated from large grain castor bean seeds. *Biochimica et Biophysica Acta*, **872**, 277–85.

Armitstead, J.G. & Ewan, P.W. (1984). Concanavalin A induced suppressor cells. *Journal of Clinical and Laboratory Immunology*, **13**, 1–10.

Asherson, G.L. & Ferluga, J. (1973). Contact sensitivity in the mouse. X. Nonspecific cytotoxicity of T blasts in the draining lymph nodes. *Immunology*, **25**, 471–83.

Ashford, D., Desai, N.N., Allen, A.K., Neuberger, A., O'Neill, M.A. & Selvendran, R.R. (1982). Structural studies of the carbohydrate moieties of lectins from potato (*Solanum tuberosum*) tubers and thorn apple (*Datura stramonium*) seeds. *The Biochemical Journal*, **201**, 199–208.

Aub, J.C., Tieslau, C. & Lankester, A. (1963). Reaction of normal and tumor cell surfaces to enzymes. I. Wheat germ lipase and associated mucopolysaccharides. *Proceedings of the National Academy of Sciences U.S.A.*, **50**, 613–19.

Aubry, M. & Boucrot, P. (1986). Etude comparée de la digestion des viciline, legumine et lectine radiomarquées de *Pisum savitum* chez le rat. *Annals of Nutrition & Metabolism*, **30**, 175–82.

Aune, T.M. & Pierce, C.W. (1981a). Mechanism of action of macrophage-derived suppressor factor produced by soluble immune response suppressor-treated macrophages. *The Journal of Immunology*, **127**, 368–72.

—(1981b). Identification and initial characterization of a nonspecific suppressor factor (macrophage S) produced by soluble immune response suppressor (SIRS)-treated macrophages. *The Journal of Immunology*, **127**, 1128–33.

Auricchio, S., de Ritis, G., de Vincenzi, M., Occorsio, P. & Silano, V. (1982). Effects of gliadin-derived peptides from bread and durum wheats on small intestine cultures from rat fetus and coeliac children. *Pediatric Research*, **16**, 1004–10.

Auricchio, S., de Ritis, G., de Vincenzi, M., Mancini, E., Minetti, M., Sapora, O. & Silano, V. (1984). Agglutinating activity of gliadin-derived peptides from bread wheat: Implications for celiac disease pathogenesis. *Biochemical and Biophysical Research Communications*, **121**, 428–33.

Auricchio, S., de Ritis, G., de Vincenzi, M. & Silano, V. (1985). Toxicity mechanisms of wheat and other cereals in celiac disease and related entheropathies. *Journal of Pediatric Gastroenterology and Nutrition*, **4**, 923–30.

Bach, M.K. & Brashler, J.R. (1975). Inhibition of IgE and compound 48/80-induced histamine release by lectins. *Immunology*, **29**, 371–86.

Baintner, K. (1986). Intestinal absorption of macromolecules and immune transmission from mother to young. *CRC Press*, Boca Raton, Florida.

Baldo, B.A., Boniface, P.A. & Simmonds, D.H. (1982*a*). Lectins as the cytochemical probes of developing wheat grain. II. Reaction of wheat germ lectin with the nucellar epidermis. *The Australian Journal of Plant Physiology*, **9**, 663–75.

—(1982*b*). Lectins as the cytochemical probes of developing wheat grain. III. Reaction of potato and wheat germ lectins with wheat protein bodies. *The Australian Journal of Plant Physiology*, **9**, 677–88.

Balin, B.J. & Broadwell, R.D. (1987). Lectin-labelled membrane is transferred to the Golgi complex in mouse pituitary cells *in vivo*. *The Journal of Histochemistry and Cytochemistry*, **35**, 489–98.

Banwell, J.G., Boldt, D.H., Meyers, J. & Weber, F.L. Jr (1983). Phytohemagglutinin derived from red kidney bean (*Phaseolus vulgaris*): A cause for intestinal malabsorption associated with bacterial overgrowth in the rat. *Gastroenterology*, **84**, 506–15.

Banwell, J.G., Abramowsky, C.R., Weber, F., Howard, R. & Boldt, D.H. (1984). Phytohemagglutinin-induced diarrheal disease. *Digestive Disease Science*, **29**, 921–9.

Banwell, J.G., Howard, R., Cooper, D. & Costerton, J.W. (1985). Intestinal microbial flora after feeding phytohemagglutinin lectins (*Phaseolus vulgaris*) to rats. *Applied Environmental Microbiology*, **50**, 68–80.

Banwell, J.G., Howard, R., Kabir, I. & Costerton, J.W. (1988). Bacterial overgrowth by indigenous microflora in the phytohemagglutinin fed rats. *Canadian Journal of Microbiology*, **34**, 1009–13.

Barbieri, L., Falasca, A., Franceschi, C., Licastro, F., Rossi, C.A. & Stirpe, F. (1983). Purification and properties of two lectins from the latex of the euphorbiaceous plants *Hura crepitans* L. (sand-box tree) and *Euphorbia characias* L. (Mediterranean spurge). *The Biochemical Journal*, **215**, 433–9.

Bardocz, S. & Hesketh, J.E. (1989). Effect of RNA synthesis inhibitors on insulin-induced protein synthesis by 3T3 cells. *International Journal of Biochemistry*, **21**, 1265–8.

Bardocz, S., Grant, G., Brown, D.S., Wallace, H.M., Ewen, S.W.B. & Pusztai, A. (1989*a*). Effect of α-difluoromethylornithine on *Phaseolus vulgaris* lectin-induced growth of the rat small intestine. *Medical Science Research*, **17**, 143–5.

Bardocz, S., Grant, G., Brown, D.S., Ewen, S.W.B. & Pusztai, A. (1989*b*). Involvement of polyamines in *Phaseolus vulgaris* lectin-induced growth of rat pancreas *in vivo*. *Medical Science Research*, **17**, 309–11.

Bardocz, S., Brown, D.S., Grant, G. & Pusztai, A. (1990*a*). Luminal and basolateral uptake by rat small intestine stimulated to grow by *Phaseolus vulgaris* phytohaemagglutinin *in vivo*. *Biochimica et Biophysica Acta*, **1034**, 46–52.

Bardocz, S., Grant, G., Brown, D.S., Ewen, S.W.B., Nevison, I. & Pusztai, A. (1990*b*). Polyamine metabolism and uptake during *Phaseolus vulgaris* lectin, PHA-induced growth of rat small intestine. *Digestion,* **46** (suppl 2) 360–6.

Barkai-Golan, R., Mirelman, D. & Sharon, N. (1978). Studies on growth inhibition by lectins of Penicillia and Aspergilli. *Archives of Microbiology*, **116**, 119–24.

Barondes, S.H. (1988). Bifunctional properties of lectins: lectins redefined. *Trends in Biochemical Sciences*, **13**, 480–2.

Barrett, D.J., Edwards, J.R., Pietrantuono, B.A. & Ayoub, E.M. (1983). Inhibition of human lymphocyte activation by wheat germ agglutinin: a model for saccharide-specific suppressor factors. *Cellular Immunology*, **81**, 287–97.

Basbaum, A.I. & Menetrey, D. (1987). Wheat germ agglutinin-apo HRP–Gold: A new retrograde tracer for light- and electronmicroscopic single- and double-label studies. *The Journal of Comparative Neurology*, **261**, 306–18.

Basham, T.Y. & Waxdal, M.J. (1975). The stimulation of immunoglobulin production in murine spleen cells by the pokeweed mitogens. *The Journal of Immunology*, **114**, 715–16.

Basten, A., Miller, J.F.A.P., Sprent, J. & Cheers, C. (1974). Cell-to-cell interaction in the immune response. X. T-cell-dependent suppression in tolerant mice. *The Journal of Experimental Medicine*, **140**, 199–217.

Becker, J.W., Reeke, G.N. Jr, Cunningham, B.A. & Edelman, G.M. (1976). New evidence on the location of the saccharide-binding site of concanavalin A. *Nature*, London, **259**, 406–9.

Becker, M.J., Drucker, I., Parkas, R., Steiner, Z. & Klajman, A. (1981). Monocyte-mediated regulation of cellular immunity in humans: loss of suppressor activity with ageing. *Clinical and Experimental Immunology*, **45**, 439–46.

Begbie, R. (1979). A non-aqueous method for the subcellular fractionation of cotyledons from dormant seeds of *Phaseolus vulgaris* L. *Planta*, **147**, 103–10.

Begbie, R. & King, T.P. (1985). The interaction of dietary lectin with porcine small intestine and production of lectin-specific antibodies. In *Lectins, Biology, Biochemistry, Clinical Biochemistry* (Bog-Hansen, T.C. & Breborovicz, J. eds), vol. 4. pp. 15–27. Walter de Gruyter, Berlin & New York.

Bender, A.E. & Reaidi, G.B. (1982). Toxicity of kidney beans (*Phaseolus vulgaris*) with particular reference to lectins. *The Journal of Plant Foods*, **4**, 15–22.

Beretta, A., Persson, U., Ramos, T. & Moller, G. (1982). Concanavalin-A inhibits the effector phase of specific cytotoxicity. *Scandinavian Journal of Immunology*, **16**, 181–9.

Berridge, M.J. (1984). Inositol triphosphate and diacylglycerol as second messengers. *The Biochemical Journal*, **220**, 345–60.

—(1987). Inositol lipids and cell proliferation. *Biochemica et Biophysica Acta*, **907**, 33–45.

Berridge, M.J., Heslop, J.P., Irvine, R.F. & Brown, K.D. (1985). Inositol lipids and cell proliferation. *The Biochemical Society Transactions*, **13**, 67–71.

Bhattacharyya, L. & Brewer, F.C. (1986). Precipitation of concanavalin A by a high mannose type glycopeptide. *Biochemical and Biophysical Research Communications*, **137**, 670–4.

Biltiger, H. & Schanebli, H.P. (1974). Binding of Con A and ricin to synaptic junction of rat brain. *Nature*, London, **249**, 370–1.

Bird, G.W.G. (1954). Phyto-agglutinins. *Acta Chirurgica Belgiae*, Suppl. 1, 33–40.

Black, C.D.V., Kroczek, R.A., Barbet, J., Weinstein, J.N. & Shevach, E.M. (1988). Induction of IL-2 receptor expression *in vivo*. Response to concanavalin A. *Cellular Immunology*, **111**, 420–32.

Bladon, T., Brasch, K., Brown, D.L. & Setterfield, G. (1988). Changes in structure and protein composition of bovine lymphocyte unclear matrix during concanavalin A-induced mitogenesis. *Biochemistry and Cell Biology*, **66**, 40–53.

Blaustein, R.O., Germann, W.J., Finkelstein, A. & Das Gupta, B.R. (1987). The N-terminal half of the heavy chain of Botulinum-type A neurotoxin forms channels in planar phospholipid bilayers. *FEBS Letters*, **226**, 115–20.

Bohlool, B.B. & Schmidt, E.I. (1974). Lectins: A possible basis for specificity in the Rhizobium–legume root module symbiosis. *Science*, **185**, 269–71.

Boldt, D.H. & Banwell, J.G. (1985). Binding of isolectins from red kidney bean (*Phaseolus vulgaris*) to purified rat brush border membranes. *Biochimica et Biophysica Acta*, **843**, 230–7.

Bollini, R. & Chrispeels, M.J. (1978). Characterisation and subcellular localization of vicilin and phytohemagglutinin, the two major reserve proteins of *Phaseolus vulgaris* L. *Planta*, **142**, 291–8.

—(1979). The rough endoplasmic reticulum is the site of reserve-protein synthesis in developing *Phaseolus vulgaris* cotyledons. *Planta*, **146**, 487–501.

Bollini, R., Vitale, A. & Chrispeels, M.J. (1983). *In vivo* and *in vitro* processing of seeds reserve protein in the endoplasmic reticulum: evidence for two glycosylation steps. *The Journal of Cell Biology*, **96**, 999–1007.

Bollini, R., Ceriotti, A., Daminati, G.M. & Vitale, A. (1985). Glycosylation is not needed for the intracellular transport of phytohemagglutinin in developing *Phaseolus vulgaris* cotyledons and for the maintenance of its biological activities. *Physiologia Plantarum*, **65**, 15–22.

Bolwell, G.P. (1987). Elicitor induction of the synthesis of a novel lectin-like arabinosylated hydroxyproline-rich glycoprotein in suspension cultures of *Phaseolus vulgaris* L. *Planta*, **172**, 184–91.

Bonavida, B. & Katz, J. (1985). Studies on the induction and expression of T cell-mediated immunity. XV. Role of non-MHC papain-sensitive target structures and Lyt-2 antigens in allogeneic and xenogeic lectin-dependent cellular cytotoxicity (LDCC). *The Journal of Immunology*, **135**, 1616–23.

Bond, H.M., Chaplin, M.F., & Bowles, D.J. (1985). Interaction of soya-bean agglutinin with purified glycoconjugates and soya-bean seed components. *The Biochemical Journal*, **228**, 127–36.

Bonnevie-Nielsen, V. (1981). Effects of caerulein and trypsin inhibitor on the endocrine mouse pancreas. *Acta Endocrinology*, **96**, 227–34.

Borberg, H., Woodruff, J., Hirschhorn, R., Gesner, B., Meischer, P.A. & Silber, R. (1966). Phytohemagglutinin: Inhibition of the agglutinating activity by *N*-acetyl-D-galactosamine. *Science*, **154**, 1019–20.

Borges, L.F. & Sidman, R.L. (1982). Axonal transport of lectins in the peripheral nervous system. *Journal of Neuroscience*, **2**, 647–53.

Borrebaeck, C.A.K. (1984). Detection and characterization of a lectin from non-seed tissue of *Phaseolus vulgaris*. *Planta*, **161**, 223–8.

Borrebaeck, C.A.K. & Schon, A. (1987). Antiproliferative response of human leukemic cells: Lectin-induced inhibition of DNA synthesis and cellular metabolism. *Cancer Research*, **47**, 4345–50.

Borrebaeck, C.A.K., Bristulf, J. & Jergil, B. (1987). Antiproliferative response of human leukemic cells. Modulation of cytosolic protein kinase C activity by phytohemagglutinin. *Cancer Letters*, **38**, 181–9.

Bowles, D.J. & Marcus, S. (1981). Characterisation of receptors for the endogenous lectins of soybean and jackbean seeds. *FEBS Letters*, **129**, 135–8.

Bowles, D.J. & Pappin, D.J. (1988). Traffic and assembly of concanavalin A. *Trends in Biochemical Sciences*, **13**, 60–4.

Bowles, D.J., Andralojc, J. & Marcus, S. (1982). Identification of an endogenous Con A binding polypeptide as the heavy subunit of α-mannosidase. *FEBS Letters*, **140**, 234–6.

Bowles, D.J., Marcus, S.E., Pappin, D.J.C., Findlay, J.B.C., Eliopoulos, E., Maycox, P.R. & Burgess, J. (1986). Posttranslational processing of concanavalin A precursors in jackbean cotyledons. *The Journal of Cell Biology*, **102**, 1284–97.

Boyd, W.C. & Reguera, R.M. (1949). Studies on haemagglutinins present in seeds of some representatives of the family leguminosae. *The Journal of Immunology*, **62**, 333–9.

Bradley, J.L., Silva, H.M. & McGuire, P.M. (1987). Depurination of yeast 26 S ribosomal RNA by recombinant ricin A chain. *Biochemical and Biophysical Research Communications*, **149**, 588–93.

Brewer, F.C., Bhattacharyya, L., Brown, R.D. & Koenig, S.H. (1985). Interactions of concanavalin A with a trimannosyl oligosaccharide fragment of complex and high mannose type glycopeptides. *Biochemical and Biophysical Research Communications*, **127**, 1066–71.

Brock, J.H. (1981). The effects of iron and transferrin on the response of serum-free cultures of mouse lymphocytes to concanavalin A and lipopolysaccharide. *Immunology*, **43**, 387–92.

Broekaert, W.F., Nsimba-Lubaki, M., Peeters, B. & Peumans, W.J. (1984). A lectin from the elder (*Sambucus nigra*) bark. *The Biochemical Journal*, **221**, 163–9.

Broekaert, W.F., Lambrechts, D., Verbelen, J.P. & Peumans, W.J. (1988). Datura stramonium agglutinin. Location in the seeds and release upon inhibition. *Plant Physiology*, **86**, 569–74.

Broglie, K.E., Gaynor, J.J. & Broglie, R.M. (1986). Ethylene-regulated gene expression: Molecular cloning of the genes encoding an endochitinase from Phaseolus vulgaris. *Proceedings of the National Academy of Sciences USA*, **83**, 6820–4.

Brown, J.W.S., Osborn, T.C., Bliss, F.A. & Hell, T.C. (1982). Bean lectins. Part 1: Relationships between agglutinating activity and electrophoretic variation in the lectin-containing G_2/albumin seed proteins of French bean (*Phaseolus vulgaris* L.). *Theoretical and Applied Genetics*, **62**, 263–71.

Brunngraber, E.C. (1969). Possible role of glycoproteins in neural function. *Perspectives in Biology and Medicine*, **12**, 467–70.

Buffard, D., Kaminski, P.A. & Strosberg, A.D. (1988). Lectin gene expression in pea (*Pisum sativum* L.) roots. *Planta*, **173**, 367–72.

Buffington, C.K., El-Shiekh, T., Kitabohi, A.E. & Matteri, R. (1986). Phytohemagglutinin (PHA) activated human T lymphocytes: concomitant appearance of insulin binding, degradation and insulin-mediated activation of pyruvate dehydrogenase (PDH). *Biochemical and Biophysical Research Communications*, **134**, 412–19.

Bulajic, M., Cuperlovic, M., Movsesijan, M. & Borojevic, D. (1986). Interaction of dietary lectin (phytohemagglutinin) with the mucosa of rat digestive tract – Immunofluorescence studies. *Periodicum Biologorum*, **38**, 331–76.

Butterworth, A.G. & Lord, J.M. (1983). Ricin and *Ricinus communis* agglutinin subunits are all derived from a single-size polypeptide precursor. *European Journal of Biochemistry*, **137**, 57–65.

Callow, J.A. (1977). Recognition, resistance and the role of plant lectins in host–parasite interactions. *Advances in Botanical Research*, **4**, 1–49.

Calne, R.Y., Wheeler, J.R. & Hurn, B.A.L. (1965). Combined immunosuppressive action of phytohaemagglutinin and azathioprine (imuran) on dogs with renal homotransplants. *British Medical Journal*, **2**, 154–5.

Cammue, B.P.A., Peters, B. & Peumans, W.J. (1986). A new lectin from (*Tulipa*) bulbs, *Planta*, **169**, 583–8.

Campbell, R.M. & Scanes, C.G. (1988). Pharmacological investigations on the lipolytic and antilipolytic effects of growth hormone (GH) in adipose tissue *in vitro*: evidence for the involvement of calcium and polyamines. *Proceedings of the Society for experimental Biology and Medicine*, **188**, 177–84.

Campenot, R.B. (1977). Local control of neurite development by nerve growth factor. *Proceedings of the National Academy of Science USA*, **74**, 4516–19.

Cantrell, D.A. & Smith, K.A. (1983). Transient expression of interleukin 2 receptors. *The Journal of Experimental Medicine*, **158**, 1895–911.

Carpene, C., Berlan, M. & Lafontan, M. (1983). Jack of functional antilipolytic alpha 2-adrenoreceptor in rat fat cell: comparison with hamster adipocyte. *Comparative Biochemistry and Physiology* [C], **74**, 41–5.

Carpenter, G. & Cohen, S. (1977). Influence of lectins on the binding of ^{125}I-labeled EGF to human fibroblasts. *Biochemical and Biophysical Research Communications*, **79**, 545–52.

Carrington, D.M., Auffret, A. & Hanke, D.R. (1985). Polypeptide ligation occurs during post-translational modification of concanavalin A. *Nature*, London, **313**, 64–7.

Casalounge, C. & Pont Lezica, R. (1985). Potato lectin: a cell wall glycoprotein. *Plant Cell Physiology*, **26**, 1533–9.

Casselas, P., Brown, J.P., Gros, O., Gros, P., Hellstrom, I., Jensen, F.K., Poncelet, P., Roncucci, R., Vidal, H. & Hellstrom, K.E. (1982). Human melanoma cells can be killed *in vitro* by an immunotoxin specific for melanoma-associated antigen p97. *International Journal of Cancer*, **30**, 437–43.

Casselas, P., Bourrie, B.J.P., Gros, P. & Jansen, F.K. (1984). Kinetics of cytotoxicity induced by immunotoxins: enhancement by lysosomotropic amines and carboxylic ionophores. *The Journal of Biological Chemistry*, **259**, 9359–64.

Castagna, M., Takai, Y., Kaibuchi, K., Sano, K., Kikkawa, U. & Nisbizuka, Y. (1982). Direct activation of calcium-activated, phospholipid-dependent protein kinase by tumor-promoting phorbol esters. *The Journal of Biological Chemistry*, **257**, 7847–51.

Castresana, M.C., Serra, M.T., Rodriguez, J.F. & Tejerina, G. (1987). Distribution of lectin during the life cycle of *Phaseolus vulgaris* L. *Plant Science*, **48**, 79–88.

Catsimpoolas, N. & Meyer, E.W. (1969). Isolation of soybean hemagglutinin and demonstration of multiple form by isoelectrics focusing. *Archives of Biochemistry and Biophysics*, **132**, 279–85.

Cawley, D.B., Hedblom, M.L. & Houston, L.L. (1978). Homology between ricin and *Ricinus communis* agglutinin: Amino terminal sequence analysis and protein synthesis inhibition studies. *Archives of Biochemistry and Biophysics*, **190**, 744–55.

Ceri, H., Falkenberg-Anderson, K., Fang, R., Costerton, J.W., Howard, R. & Banwell, J.G. (1988). Bacteria-lectin interactions in phytohemagglutinin-induced bacterial overgrowth of the small intestine. *Canadian Journal of Microbiology*, **34**, 1003–8.

Chakraborty, N.G., Bose, S.R. & Chowdhury, J.R. (1987). Changes induced by transplantable tumour on the surface of lymphocytes of mice as evidenced by lectin mediated agglutination. *Medical Science Research*, **15**, 893–4.

Chappel, J. & Chrispeels, M.J. (1986). Transcriptional and posttranscriptional control of phaseolin and phytohemagglutinin gene expression in developing cotyledons of *Phaseolus vulgaris*. *Plant Physiology*, **81**, 50–4.

Chatamara, K., Daniel, P.M., Kendall, M.D. & Lam, D.K.C. (1985). Atrophy of the thymus in rats rendered diabetic by streptozotocin. *Hormones and Metabolic Research*, **17**, 630–2.

Chatterjee, R., Guha, S. & Chattopadhyay, J. (1986). Concanavalin A-induced agglutination of Ehrlich ascites carcinoma cells during growth of the tumor in mice. *Neoplasma*, **33**, 685–9.

Chen, A.P.T. & Phillips, D.A. (1976). Attachment of Rhizobium to legume roots as the basis for specific interactions. *Physiologia Plantarum*, **38**, 83–8.

Chiang, C.M. & Weiner, N. (1987). Gastrointestinal uptake of liposomes II. In vivo studies. *International Journal of Pharmaceutics*, **40**, 143–50.

Chilson, O.P., Boylston, A.W. & Crumpton, M.J. (1984). *Phaseolus vulgaris* phytohemagglutinin (PHA) binds to the human T lymphocyte antigen receptor. *The RMBO Journal*, **3**, 3239–45.

Chou, P.Y. & Fasman, G.D. (1978). Empirical predictions of protein conformation. *Annual Review of Biochemistry*, **47**, 251–76.

Chrispeels, M.J. (1983*a*). Incorporation of fucose into the carbohydrate moiety of phytohemagglutinin in developing *Phaseolus vulgaris* cotyledons. *Planta*, **157**, 454–61.

—(1983*b*). The Golgi apparatus mediates the transport of phytohemagglutinin to the protein bodies in bean cotyledons. *Planta*, **158**, 140–51.

—(1984). Biosynthesis, processing and transport of storage proteins and lectins in the cotyledons of developing legume seeds. *Philosophical Transactions of the Royal Society, London Series B*, **304**, 309–22.

Chrispeels, M.J. & Bollini, R. (1982). Characteristics of membrane-bound lectin in developing *Phaseolus vulgaris* cotyledons. *Plant Physiology*, **70**, 1425–8.

Chrispeels, M.J. & Greenwood, J.S. (1987). Heat stress enhances phytohemagglutinin synthesis but inhibits its transport out of the endoplasmic reticulum. *Plant Physiology*, **83**, 778–84.

Chrispeels, M.J., Higgins, T.J.V., Craig, S. & Spencer, D. (1982). The role of the endoplasmic reticulum in the synthesis of reserve proteins and the kinetics of their transport to protein bodies in developing pea cotyledons. *The Journal of Cell Biology*, **93**, 5–14.

Clarke, A.E. & Knox, R.B. (1979). Plant and immunity. *Developmental and Comparative Immunology*, **3**, 571–89.

Clarke, A.E., Anderson, R.L. & Stone, B.A. (1979). Form and function of arabinogalactans and arabinogalactan proteins. *Phytochemistry*, **18**, 521–40.

Clevers, H.C., de Bresser, A., Kleinveld, H., Gmelig-Meyling, F.H. & Ballieux, R.E. (1986). Wheat germ agglutinin activates human T lymphocytes by stimulation of phosphoinositide hydrolysis. *The Journal of Immunology*, **136**, 3180–3.

Cockroft, S. (1987). Phosphoinositide phosphodiesterase: regulation by a novel guanine nucleotide binding protein, G_p. *Trends in Biochemical Sciences*, **12**, 75–8.

Cody, R. & Wicha, M. (1986). Clustering of cell surface laminin enhances its association with cytoskeleton. *Experimental Cell Research*, **165**, 107–16.

Concon, J.M., Newburg, D.S. & Eades, S.N. (1983). Lectins in wheat gluten proteins. *The Journal of Agriculture and Food Chemistry*, **31**, 939–41.

Courtneidge, S.A. & Heber, A. (1987). An 81kD protein complexed with middle T antigen and pp60^{c-mrc}: a possible phosphatidylinositol kinase. *Cell*, **50**, 1031–7.

Crossland, W.J. (1985). Anterograde and retrograde axonal transport of native and derivatized wheat germ agglutinin in the visual system of the chicken. *Brain Research*, **347**, 11–27.

Crowley, J.F., Goldstein, I.J., Arnarp, J. & Lonngren, J. (1984). Carbohydrate binding studies on the lectin from *Datura stramonium* seeds. *Archives of Biochemistry and Biophysics*, **231**, 524–33.

Cuatrecasas, P. (1973). Interaction of concanavalin A and wheat germ agglutinin with the insulin receptor of fat cells and liver. *The Journal of Biological Chemistry*, **248**, 3528–34.

Cuatrecasas, P. & Tell, G.P.E. (1973). Insulin-like activity of concanavalin A and wheat germ agglutinin: direct interactions with insulin receptors. *Proceedings of the National Academy of Sciences, USA*, **70**, 485–9.

Cummings, R.D. & Kornfeld, S. (1982). Characterization of the structural determinants required for the high affinity interaction of the asparaginine-linked oligosaccharides with immobilized *Phaseolus vulgarus* leukoagglutinating and erythroagglutinating lectins. *The Journal of Biological Chemistry*, **257**, 11230–4.

Cunningham, B.A., Hemperly, J., Hopp, T.P. & Edelman, G.M. (1979). Favin versus concanavalin A: Circularly permuted amino acid sequences. *Proceedings of the National Academy of Sciences, USA*, **76**, 3218–22.

Cutler, R.L., Metcalf, D., Nicola, N.A. & Johnson, G.R. (1985). Purification of a multipotential colony-stimulating factor from pokeweed mitogen-stimulated mouse spleen cell-conditioned medium. *The Journal of Biological Chemistry*, **260**, 6579–81.

Dabauvalle, M.C., Schulz, B., Scheer, U. & Peters, R. (1988). Inhibition of nuclear accumulation of karyophylic proteins in living cells by micro-injection of the lectin wheat germ agglutinin. *Experimental Cell Research*, **174**, 291–6.

Dallas, W.S. & Falkow, S. (1980). Amino acid sequence homology between cholera toxin and *E. coli* heat-labile toxin. *Nature*, London, **288**, 499–501.

Davidson, L.A. & Lonnerdal, B. (1988). Specific binding of lactoferrin to brush-border membrane: ontogeny and effect of glycan chain. *The American Journal of Physiology*, **254**, G580–5.

Davila, D.R., Franklin, R.A., Kleiss, A.J. & Kelley, K.W. (1987). A soluble 61-kD protein is associated with inhibition of lectin-induced proliferation and IL-2 synthesis. *Proceedings of the Society for Experimental Biology and Medicine*, **186**, 1–12.

Davis, L.I. & Blobel, G. (1986). Identification and characterization of a nuclear pore complex protein. *Cell*, **45**, 699–709.

Dazzo, F.B. & Brill, W.J. (1979). Bacterial polysaccharide which binds to *Rhisobium trifolii* to clover root hairs. *The Journal of Bacterology*, **137**, 1362–73.

Dazzo, F.B. & Truchet, G.L. (1983). Interactions of lectins and their saccharide receptors in the *Rhizobium*-legume symbiosis. *The Journal of Membrane Biology*, **73**, 1–16.

Dazzo, F.B., Yanke, W.E. & Brill, W.J. (1978). Trifoliin: A *Rhizobium* recognition protein from white clover. *Biochimica et Biophysica Acta*, **539**, 276–86.

de Aizpurua, H.J. & Russel-Jones, G.J. (1988). Oral vaccination. Identification of classes of proteins that provoke an immune response on oral feeding. *The Journal of Experimental Medicine*, **167**, 440–51.

Dean, J.W.III, Chandrasekaran, S. & Tanzer, M.L. (1988). Lectins inhibit cell binding and spreading on a laminin substrate. *Biochemical and Biophysical Research Communications*, **156**, 411–16.

DeGeorge, J.J. & Carbonetto, S. (1986). Wheat germ agglutinin inhibits nerve fiber growth and concanavalin A stimulates nerve fiber initiation in culture of dorsal root ganglia neurons. *Developmental Brain Research*, **28**, 169–75.

Desai, N.N. & Allen, A.K. (1979). The purification of potato lectin by affinity chromatography on an *N,N',N''*-triacetylchitotriose–Sepharose matrix. *Analytical Biochemistry*, **93**, 88–90.

Dey, P.M., Naik, S. & Pridham, J.B. (1982*a*). The lectin nature of α-galactosidases from *Vicia faba* seeds. *FEBS Letters*, **150**, 233–7.

Dey, P.M., Pridham, J.B. & Sumar, N. (1982*b*). Multiple forms of *Vicia faba* α-galactosidases and their relationships. *Phytochemistry*, **21**, 2195–9.

Dey, P.M., Naik, S. & Pridham, J.B. (1986). Properties of α-galactosidase II^2 from *Vicia faba* seeds. *Planta*, **167**, 114–18.

Dhaunsi, G.S., Garg, U.C., Sidhu, G.S. & Bhatnagar, R. (1985). Enzymic and transport studies on the rat intestinal brush border membrane vesicles bound to pea and lentil lectins. *IRCS Medical Science*, **13**, 469–70.

Diaz, L.C., Lems-van Kan, P., Van der Schaal, I.A.M. & Kijne, J.W. (1984). Determination of pea (*Pisum sativum* L.) root lectin using an enzyme-linked immunoassay. *Planta*, **161**, 302–7.

Diaz, C.L., van Spronsen, P.C., Bakhuizen, R., Logman, G.J.J., Lugtenberg, E.J.J. & Kijne, J.W. (1986). Correlation between infection by *Rhizobium leuminosarum* and lectin on the surface of *Pisum sativum* roots. *Planta*, **168**, 350–9.

Diaz, L.C., Melchers, L.S., Hooykaas, P.J.J., Lugtenberg, B.J.J. & Kijne, J.W. (1989). Root lectin as a determinant of host–plant specificity in the *Rhizobium*–legume symbiosis. *Nature*, London, **338**, 579–81.

Dillner-Centerlind, M.L., Hellstrom, U., Robertsson, E.S., Hammarstrom, S. & Perlmann, P. (1980). Mitogenic responsiveness of human T-lymphocyte subpopulations: regulation by suppressive Fc-receptor-bearing T cells and influence of fractionation procedures. *Scandinavian Journal of Immunology*, **12**, 13–21.

Dixon, H.B.F. (1981). Defining a lectin. *Nature*, London, **292**, 192.

Donatucci, D.A., Liener, I.E. & Gross, C.J. (1987). Binding of navy bean (*Phaseolus vulgaris*) lectin to the intestinal cells of the rat and its effect on the absorption of glucose. *Journal of Nutrition*, **117**, 2154–60.

Dorland, L., van Halbeck, H., Vliegenthart, J.F.G., Lis, H. & Sharon, N. (1981). Primary structure of the carbohydrate chain of soybean agglutinin. A reinvestigation by high resolution ^{1}H NMR spectroscopy. *The Journal of Biological Chemistry*, **256**, 7708–11.

Doyle, R. & Keller, K. (1984). Lectins in diagnostic microbiology. *European Journal of Clinical Microbiology*, **3**, 4–9.

Draaijer, M., Koninkx, J., Hendriks, H., Kirk, M., van Dijik, J. & Mouwen, J.I. (1989). Actin cytoskeletal lesions in differentiated human colon carcinoma Caco-2 cells after exposure to soyabean agglutinins. *Biology of the Cell*, **65**, 29–35.

Draetta, G., Piwnica-Worms, H., Morrison, D., Draker, B., Roberts, T. & Beach, D. (1988). Human *cdc2* protein kinase is a major cell-cycle regulated tyrosine kinase substrate. *Nature*, London, **336**, 738–44.

Drobniewski, F.A. (1989). Immunotoxins up to the present day. *Bioscience Reports*, **9**, 139–56.

Dumas, M., Schwab, M.E. & Thoenen, H. (1979). Retrograde axonal transport of specific macromolecules as a tool for characterizing nerve terminal membranes. *Journal of Neurobiology*, **10**, 179–97.

Dumont, M.E. & Richards, F.M. (1988). The pH-dependent conformational change of diphtheria toxin. *The Journal of Biological Chemistry*, **263**, 2087–97.

Dutton, R.W. (1972). Inhibitory and stimulatory effects of concanavalin A on the response of mouse spleen cell suspensions to antigen. I. Characterization of the inhibitory cell activity. *The Journal of Experimental Medicine*, **136**, 1445–60.

—(1975). Suppressor T cells. *Transplantation Reviews*, **26**, 39–55.

Eckhardt, A.E., Malone, B.N. & Goldstein, I.J. (1982). Inhibition of Erlich ascites tumor cell growth by *Griffonia simplicifolia* I lectin *in vivo*. *Cancer Research*, **42**, 2977–9.

Edelman, G.M. (1983). Cell adhesion molecules. *Science*, **219**, 450–7.

Edelman, G.M. & Wang, J.L. (1978). Binding and functional properties of concanavalin A and its derivatives. III. Interactions with indoleacetic acid and other hydrophobic ligands. *The Journal of Biological Chemistry*, **253**, 3016–22.

Edelman, G.M., Cunningham, B.A., Reeke, G.N. Jr, Becker, J.W., Waxdal, M.J. & Wang, J.L. (1972). Covalent and three-dimensional structure of concanavalin A. *Proceedings of the National Academy of Sciences, USA*, **62**, 2580–5.

Einhoff, W. & Rüdiger, H. (1986*a*). Isolation of the *Canavalia ensiformis* seed α-mannosidase by chromatography on concanavalin A, the lectin from the same plant, without involving its sugar binding site. *Biological Chemistry. Hoppe-Seyler*, **367**, 313–20.

—(1986*b*). Interaction of the α-mannosidase from *Canavalia ensiformis* with the lectin from the same plant, concanavalin A. *Biological Chemistry. Hoppe-Seyler*, **367**, 943–9.

Elfstrand, M. (1898). Über blutkorperchenagglutinierende Eiweisse. In *Görbersdorfer Veröffentlichungen* a.Band I (R. Kobert, ed.), pp. 1–159. Enke, Stuttgart.

Endo, Y. & Tsurugi, K. (1987). *N*-glycosidase activity of ricin A-chain. Mechanism of action of the toxic lectin ricin on eukaryotic ribosomes. *The Journal of Biological Chemistry*, **262**, 8128–30.

Erikson, R.H., Kim, J., Sleisenger, M.H. & Kim, Y.S. (1985). Effects of lectins on the activity of brush border membrane-bound enzymes of rat small intestine. *Journal of Pediatric Gastroenterology and Nutrition*, **4**, 984–91.

Esquerre-Tugaye, M.T. & Maxau, D. (1974). Effect of a fungal disease on extensin, the plant cell wall glycoprotein. *Journal of Experimental Botany*, **25**, 509–13.

Esquivel, P., Mena, M. & Folch, H. (1982). Suppression of autoimmune thyroiditis by phytohemagglutinin. *Cellular Immunology*, **67**, 410–13.

Etzler, M.E. (1985). Plant lectins: Molecular and biological aspects. *Annual Review of Plant Physiology*, **36**, 209–34.

—(1986). Distribution and function of plant lectins. In *The Lectins* (Liener, I.E., Sharon, N. & Goldstein, I.J. eds.), pp. 371–435. Academic Press, Orlando.

Etzler, M.E. & Borrebaeck, C. (1980). Carbohydrate binding activity of a lectin-like glycoprotein from stems and leaves of *Dolichos biflorus*. *Biochemical and Biophysical Research Communications*, **96**, 92–7.

Etzler, M.E. & Branstrator, M.L. (1974). Differential localization of cell surface and secretory components in rat intestinal epithelium by use of lectins. *Journal of Cell Biology*, **62**, 329–43.

Etzler, M.E. & Kabat, E.A. (1970). Purification and characterization of a lectin (plant hemagglutinin) with blood group A specificity from *Dolichos biflorus*. *Biochemistry*, **9**, 869–77.

Etzler, M.E., MacMillan, S., Scates, S., Gibson, D.M., James, D.W. Jr, Cole, D. & Thayer, S. (1984). Subcellular localizations of two *Dolichos biflorus* lectins. *Plant Physiology*, **76**, 871–8.

Fabian, R.H. & Coulter, J.D. (1985). Transneuronal transport of lectins. *Brain Research*, **344**, 41–8.

Falasca, A., Franceschi, C., Rossi, C.A. & Stirpe, F. (1980). Mitogenic and haemagglutinating properties of a lectin purified from *Hura crepitans* seeds. *Biochimica et Biophysica Acta*, **632**, 95–105.

Faltynek, C.R., Princler, G.L., Ruscetti, F.W. & Birchenal-Sparks, M. (1988). Lectins modulate the internalization of recombinant interferon-α A and induction of 2′,5′-Oligo(A) synthetase. *The Journal of Biological Chemistry*, **263**, 7112–17.

Farkas, R., Manor, Y. & Klajman, A. (1986). Generation of B suppressor cells by phytohaemagglutinin. *Immunology*, **57**, 395–8.

Farrar, W.L. & Anderson, W.B. (1985). Interleukin-2 stimulates association of protein kinase C with plasma membrane. *Nature*, London, **315**, 233–5.

Farrar, W.L., Thomas, T.P. & Anderson, W.B. (1985). Altered cytosol/membrane enzyme redistribution of interleukin-3 activation of protein kinase C. *Nature*, London, **315**, 235–7.

Felsted, R.L., Pokrywka, G., Chen, C., Egorin, M.J. & Bachur, N.R. (1982). Radioimmunoassay and immunochemistry of *Phaseolus vulgaris* phytohemagglutinin: verification of isolectin subunit structures. *Archives of Biochemistry and Biophysics*, **215**, 89–99.

Finlay, D.R., Newmeyer, D.D., Price, T.M. & Forbes, D.J. (1987). Inhibition of *in vitro* nuclear transport by a lectin that binds to nuclear pores. *The Journal of Cell Biology*, **104**, 189–200.

Fleischer, T.A., Greene, W.C., Blaese, R.M. & Waldman, T.A. (1981). Soluble suppressor supernatants elaborated by concanavalin A-activated human mononuclear cells. II. Characterization of a soluble suppressor of B cell immunoglubulin production. *The Journal of Immunology*, **126**, 1192–70.

Fodstad, O., Kvalheim, G., Gedal, A., Lotsberg, J., Aamdal, S., Host, H. & Pihl, A. (1984). Phase 1 study of the plant protein ricin. *Cancer Reaearch*, **44**, 862–5.

Foriers, A., Lebrun, E., Van Rapenbush, R., de Neve, R. & Strosberg, A.D. (1981). The structure of the lentil (*Lens culinaris*) lectin. Amino acid sequence determination and prediction of the secondary structure. *The Journal of Biological Chemistry*, **256**, 5550–60.

Fountain, D.W., Foard, D.E., Replogle, W.D. & Yang, W.K. (1977). Lectin release by soybean seeds. *Science*, **197**, 1185–7.

Fox, E.J., Cook, R.G., Lewis, D.E. & Rich, R.R. (1986). Proliferative signals for suppressor T cells. Helper cells stimulated with pokeweed mitogen *in vitro* produce a suppressor cell growth factor. *The Journal of Clinical Investigation*, **78**, 214–20.

Foxwell, B.M.J., Donovan, T.A., Thorpe, P.E. & Wilson, G. (1985). The removal of carbohydrates from ricin with endoglycosidases H, F and D and α-mannosidase. *Biochimica et Biophysica Acta*, **840**, 193–203.

Foxwell, B.M.J., Blakey, D.C., Brown, A.N.F., Donovan, R.A. & Thorpe, P.E. (1987). The preparation of deglycosylated ricin by recombination of glycosidase-treated A and B-chains: effects of deglycosylation on toxicity and *in vivo* distribution. *Biochimica et Biophysica Acta*, **923**, 59–65.

Freed, D.L.J. (1979). Dietary lectins and the anti-nutritive effect of gut allergy. In *Protein Transmission through Living Membrances* (Hemmings, V.A., ed.), pp. 411–22. Elsevier/North Holland, Amsterdam.

Freed, D.L.J. & Buckley, C.H. (1979). Mucotractive effect of lectin. *Lancet*, i, 585–6.

Froelich, C.J., Burkett, J.S., Guiffant, S., Kingsland, R. & Brauner, D. (1988). Phytohemagglutinin induced proliferation by aged lymphocytes: Reduced expression of high affinity interleukin-2 receptors and interleukin-2 secretion. *Life Sciences*, **43**, 1583–90.

Frost, R.G., Reitherman, R.W., Miller, A.L. & O'Brien, J.S. (1975). Purification of *Ulex europeus* hemagglutinin I by affinity chromatography. *Analytical Biochemistry*, **69**, 170–9.

Fulton, R.J., Blakey, D.C., Knowles, P.P., Uhr, J.W., Thorpe P.E. & Vitetta, E.S. (1986). Purification of ricin A_1, A_2 and B-chains and characterization of their toxicity. *The Journal of Biological Chemistry*, **261**, 5314–19.

Funatsu, G., Kimjura, M. & Funatsu, M. (1979). Biochemical studies on ricin. Part XXVI. Primary structure of Ala-chain of ricin D. *Agricultural and Biological Chemistry*, **43**, 2221–4.

Funatsu, G., Yamasaki, N. & Kakinchi, S. (1987). Involvement of the B chain C terminal region in the high-affinity saccharide binding site of ricin D. *Agricultural and Biological Chemistry*, **51**, 1225–6.

Gade, W., Jack, M.A., Dahl, J.B., Schmidt, E.L. & Wold, F. (1981). The isolation and characterization of a root lectin from soybean (*Glycine max* (L) cultivar Chippewa). *The Journal of Biological Chemistry*, **256**, 12905–10.

Gade, W., Schmidt, E.L. & Wold, F. (1983). Evidence for the existence of an intracellular root-lectin in soybeans. *Planta*, **158**, 108–10.

Gallily, R., Stain, I. & Zaady, O. (1986). Dual effects of lectins on macrophages: potentiation of bacterial uptake and of bacterial activity. *Immunology Letters*, **13**, 151–8.

Galun, M., Braun, A., Frensdorff, A. & Galun, E. (1976). Hyphal walls of isolated lichen fungi. Autoradiographic localization of precursor incorporation and binding of fluorescein-conjugated lectins. *Archives of Microbiology*, **108**, 9–16.

Gansera, R., Schurz, H. & Rüdiger, H. (1979). Lectin-associated proteins from the seeds of Leguminosae. *Hoppe-Seyler's Zuricher Physiologisches Chemie*, **360**, 1579–85.

Garlick, P.J., McNurlan, M.A. & Preedy, V.R. (1980). A rapid and convenient technique for measuring the rate of protein synthesis in tissues by injection of [^{3}H] phenylalanine. *The Biochemical Journal*, **192**, 719–23.

Garnier, J., Osguthorpe, D.J. & Robson, B. (1978). Analysis of the accuracy and implications of simple methods for predicting the secondary structure of globular proteins. *Journal of Molecular Biology*, **120**, 97–120.

Gatehouse, A.M.R., Dewey, F.H., Dove, J., Fenton, K.A. & Pusztai, A. (1984). Effect of seed lectins from *Phaseolus vulgaris* on the development of larvae of *Callosobruchus maculatus*: mechanism of toxicity. *The Journal of the Science of Food and Agriculture*, **35**, 373–80.

Gatehouse, A.M.R., Shackley, S.J., Fenton, K.A., Bryden, J. & Pusztai, A. (1989). Mechanism of seed lectin tolerance by a major insect storage pest of *Phaseolus vulgaris, Acanthoscelides obtectus*. *The Journal of the Science of Food and Agriculture*, **40**, 269–80.

Gatehouse, J.A. & Boulter, D. (1980). Isolation and properties of a lectin from the roots of *Pisum sativum* (Garden pea). *Physiologia Plantarum*, **49**, 437–42.

Gebauer, G., Schiltz, E., Schimpl, A. & Rüdiger, H. (1979). Purification and characterization of a mitogenic lectin and a lectin-binding protein from *Vicia sativa*. *Hoppe-Seyler's Zuricher Physiologisches Chemie*, **360**, 1727–35.

Gebauer, G., Schimpl, A. & Rüdiger, H. (1982). Lectin-binding proteins as potent mitogens for B-lymphocytes from nu/nu mice. *European Journal of Immunology*, **12**, 491–5.

Gelfand, E.W., Cheung, R.K., Mills, G.B. & Grinstein, S. (1985). Mitogens trigger a calcium-independent signal for proliferation in phorbol ester-treated lymphocytes. *Nature*, London, **315**, 419–20.

Gerace, L., Ottaviano, Y. & Kondor-Koch, C. (1982). Identification of a major polypeptide of the nuclear pore complex. *Journal of Cell Biology*, **95**, 826–37.

Gibbons, R.J. & Dankers, I. (1981). Lectin-like constituents of food which react with components of serum, saliva and Strep. mutans. *Applied Environmental Microbiology*, **41**, 880–8.

—(1982). Inhibition of lectin-binding to saliva-treated hydroxyapatite, to buccal epithelial cells and to erythrocytes by salivary components. *American Journal of Clinical Nutrition*, **36**, 276–83.

—(1983). Association of food lectins with human oral epithelial cells *in vivo*. *Archives of Oral Biology*, **28**, 561–6.

Gibson, D.M., Stack, S., Krell, K. & House, J. (1982). A comparison of soybean agglutinin in cultivars resistant and susceptible to *Phytophthora megasperma* var. sojae (Race 1). *Plant Physiology*, **70**, 560–6.

Gilbert, M. & Fulton, A. (1985). The specificity and stability of the triton-extracted cytoskeletal framework of gerbil fibroma cells. *The Journal of Cell Science*, **73**, 335–45.

Gill, D.M. (1978). Seven toxic peptides that cross cell membranes. In *Bacterial Toxins and Cell Membranes* (Jeljazewicz, J. & Wadstrom, T., eds), pp. 291–332. Academic Press, New York.

Gleeson, P.A., Jermyn, M.A. & Clarke, A.E. (1979). Isolation of an arabinogalactan protein by lectin affinity chromatography on tridacnin-Sepharose 4B. *Analytical Biochemistry*, **92**, 41–5.

Goldberg, R.B., Hoschek, G. & Vodkin, L.O. (1983). An insertion sequence blocks the expression of a soybean lectin gene. *Cell*, **33**, 465–75.

Goldsmith, M.A. & Weiss, A. (1988). Early signal transduction by the antigen receptor without commitment to T cell activation. *Science*, **240**, 1029–31.

Goldstein, I.J. & Hayes, C.E. (1978). The lectins: Carbohydrate-binding proteins of plants and animals. *Advances in Carbohydrate Chemistry and Biochemistry*, **35**, 127–334.

Goldstein, I.J. & Poretz, R.D. (1986). Isolation, physicochemical characterization, and carbohydrate-binding specificity of lectins. In *The Lectins* (Liener, I.E., Sharon, N. & Goldstein, I.J., eds), pp. 33–247. Academic Press, Orlando.

Goldstein, I.J., Hughes, R.C., Monsigny, M., Osawa, T. & Sharon, N. (1980). What should be called a lectin. *Nature*, London, **285**, 66.

Gonzalez, J.E. & Wisnieski, B.J. (1988). An endosomal model for rapid triggering of diphtheria toxin translocation. *The Journal of Biological Chemistry*, **263**, 15257–9.

Gordon, J.A., Blumberg, S., Lis, H. & Sharon, N. (1972). Purification of soybean agglutinin by affinity chromatography on Sepharose-*N*-ε-aminocaproyl-β-D-galactopyranosylamine. *FEBS Letters*, **24**, 193–6.

Granelli-Piperno, A., Andrus, L. & Steinman, R.M. (1986). Lymphokine and nonlymphokine mRNA levels in stimulated human T cells. Kinetics, mitogen requirements, and effects of cyclosporin A. *The Journal of Experimental Medicine*, **163**, 922–37.

Grant, G., More, L.J., McKenzie, N.H. & Pusztai, A. (1982). The effect of heating on the hemagglutinating activity and nutritional properties of bean (*Phaseolus vulgaris*) seeds. *The Journal of Science of Food and Agriculture*, **33**, 1324–6.

Grant, G., More, L.J., McKenzie, N.H., Stewart, J.C. & Pusztai, A. (1983). A survey of the nutritional and haemagglutination properties of legume seeds generally available in the U.K. *The British Journal of Nutrition*, **50**, 207–14.

Grant, G., Greer, F., McKenzie, N.H. & Pusztai, A. (1985). The nutritional response of mature rats to kidney bean (*Phaseolus vulgaris*) lectins. *The Journal of the Science of Food and Agriculture*, **36**, 409–14.

Grant, B., McKenzie, N.H., Watt, W.B., Stewart, J.C., Dorward, P.M. & Pusztai, A. (1986). Nutritional evaluation of soyabean (*Glycine max*). 1. Nitrogen balance and fractionation studies. *The Journal of the Science of Food and Agriculture*, **37**, 1001–10.

Grant, G., Oliveira, J.T.A. de, Dorward, P.M., Annand, M.G., Waldron, M. & Pusztai, A. (1987a). Metabolic and hormonal changes in rats resulting from consumption of kidney bean (*Phaseolus vulgaris*) or soyabean (*Glycine max*). *Nutrition Reports International*, **36**, 763–72.

Grant, G., Watt, W.B., Stewart, J.C. & Pusztai, A. (1987b). *Effect of dietary soyabean (Glycine max) lectin and trypsin inhibitors upon the pancreas of rats. Medical Science Research*, **15**, 1197–8.

—(1987c). Changes in the small intestine and hind leg muscles of rats induced by dietary soyabean (*Glycine max*) proteins. *Medical Science Research*, **15**, 1355–6.

Grant, G., Bardocz, S., Brown, D.S. & Pusztai, A. (1989). Polyamine metabolism during pancreas enlargement induced by dietary soya-bean proteins. *Biochemical Society Transactions*, **17**, 527–8.

Grant, N.J., Oriol-Audit, C. & Dickens, M.J. (1983). Supramolecular forms of actin induced by polyamines: an electron microscopic study. *European Journal of Cell Biology*, **30**, 67–73.

Green, W.R. (1982). Studies on the mechanism of lectin-dependent T cell-mediated cytolysis: use of Lens culinaris hemagglutinin to define the role of lectin. *Advances in Experimental and Medical Biology*, **146**, 81–100.

Green, B.D. & Baenziger, J.U. (1987). Oligosaccharide specificities of *Phaseolus vulgaris* leukoagglutinating and erythroagglutinating phytohemagglutinins. Interactions with *N*-glycanase-released oligosaccharides. *The Journal of Biological Chemistry*, **262**, 12018–29.

Green, G.M. & Lyman, R.L. (1972). Feedback regulation of prokaryotic enzyme secretion as a mechanism for trypsin inhibitor induced hypersecretion in rats. *Proceedings of the Society of Experimental Biology and Medicine*, **146**, 6–12.

Greenaway, P.J. & LeVine, D. (1973). Binding of N-acetyl-neuraminic acid by wheat germ agglutinin. *Nature (New Biology)*, **241**, 191–2.

Greene, W.C., Fleischer, T.A. & Waldman, T.A. (1981). Soluble supernatants elaborated by concanavalin A-activated human mononuclear cells. I. Characterization of a soluble suppressor of T cell proliferation. *The Journal of Immunology*, **126**, 1185–91.

Greenwood, J.S., Stinissen, H.M., Peumans, W.J. & Chrispeels, M.J. (1986). *Sambucus nigra* agglutinin is located in protein bodies in the phloem parenchyma of the bark. *Planta*, **167**, 275–8.

Greer, F. & Pusztai, A. (1985). Toxicity of kidney bean (*Phaseolus vulgaris*) in rats: Changes in intestinal permeability. *Digestion*, **32**, 42–46.

Greer, F., Brewer, A.C. & Pusztai, A. (1985). Effect of kidney bean (*Phaseolus vulgaris*) toxin on tissue weight and composition and some metabolic functions of rats. *The British Journal of Nutrition*, **54**, 95–103.

Griebel, C. (1950). Erkrankungen durch Bohnenflocken (*Phaseolus vulgaris* L.) und Platterbsen (*Lathyrus tingitanus* L.). *Zeitschrift für Lebensmittel-Untersuchung und - Forschung*, **90**, 191–7.

Grinstein, S., Smith, J.D., Rowatt, C. & Dixons, S.J. (1987). Mechanism of activation of lymphocyte Na^+/H^+ exchange by concanavalin A. A calcium and protein kinase C-independent pathway. *The Journal of Biological Chemistry*, **262**, 15277–84.

Grove, D.S. & Mastro, A.M. (1987). Changes in protein kinase C and cAMP-dependent kinase in lymphocytes after treatment with 12-O-tetradecanoylphorbol-13-acetate or concanavalin A: quantitation of activities with an *in situ* gel assay. *The Journal of Cellular Physiology*, **132**, 415–27.

Gulbenkian, A., Myers, J., Egan, R.W. & Seigel, M.I. (1987). The role of a Ca^{++}/Calmodulin dependent plasma membrane Ca^{++} channel during concanavalin A activation of MC9 mast cells. *Agents and Actions*, **22**, 16–23.

Gunther, G.R., Wang, J.L., Yahara, I., Cunningham, B.A. & Edelman, G. (1973). Concanavalin A derivatives with altered biological activities. *Proceedings of the National Academy of Sciences*, USA, **70**, 1012–16.

Haass, D., Frey, R., Thiesen, M. & Kauss, H. (1981). Partial purification of a hemagglutinin associated with cell walls from hypocotyls of *Vigna radiata*. *Planta*, **151**, 490–6.

Haidvogl, M., Fritsch, G. & Grubbauer, H.M. (1979). Vergiftung durch rohe Gartenbohnen (*Phaseolus vulgaris* und *Phaseolus coccineus*) im Kindesalter. *Pädiatrie und Pädologie*, **14**, 293–6.

Halling, K.C., Halling, A.C., Murray, E.E., Ladin, B.F., Houston, L.L. & Weaver, R.F. (1985). Genomic cloning and characterization of a ricin gene from *Ricinus communis*. *Nucleic Acid Research*, **13**, 8019–33.

Halverson, L.J. & Stacey, G. (1984). Host recognition in the *Rhizobium* soybean symbiosis: detection of a protection factor in soybean root exudate which is involved in the nodulation process. *Plant Physiology*, **74**, 84–9.

—(1985). Host recognition in the *Rhizobium*-soybean symbiosis: evidence for the involvement of lectin in the nodulation. *Plant Physiology*, **77**, 621–5.

—(1986a). Signal exchange in plant-microbe interactions. *Microbiological Reviews*, **50**, 193–225.

—(1986b). Effect of lectin on nodulation by wild-type *Bradyrhizobium japonicum* and a nodulation-defective mutant. *Applied Environmental Microbiology*, **51**, 753–60.

Halvorsen, R., Gaudernack, G., Leivestad, T., Vartdal, F. & Thorsby, E. (1987). Activation of resting, pure CD4+ and CD8+ cells via CD3: Requirements for second signals. *Scandinavian Journal of Immunology*, **26**, 197–205.

Halvorsen, R., Leivestad, T., Gaudernack, G. & Thorsby, E. (1988). Role of accessory cells in the activation of pure T cells via the T cell receptor-CD3 complex or with phytohaemagglutinin. *Scandinavian Journal of Immunology*, **27**, 555–63.

Hamblin, J. & Kent, S.P. (1973). Possible role of phytohaemagglutinin in *Phaseolus vulgaris* L. *Nature (New Biology)*, **245**, 28–30.

Hammarström, S., Murphy, L.A., Goldstein, I.J. & Etzler, M.E. (1977). Carbohydrate binding specificity of four *N*-acetyl-D-galactosamine-'specific' lectins: *Helix pomatia* A hemagglutinin, soy bean agglutinin, lima bean lectin and *Dolichos biflorus* lectin. *Biochemistry*, **16**, 2750–5.

Hammarström, S., Hammarström, M.L., Sundblad, G., Arnarp, J. & Lonngren, J. (1982). Mitogenic leukoagglutinin from *Phaseolus vulgaris* binds to a pentasaccharide unit in *N*-acetyllactosamine-type glycoprotein glycans. *Proceedings of the National Academy of Sciences, USA*, **79**, 611–15.

Hammerschlag, R., Stone, G.C., Bolen, F.A., Lindsey, J.D.R. & Rlisman, M.H. (1982). Evidence that all newly synthesized proteins destined for fast axonal transport pass through the Golgi apparatus. *The Journal of Cell Biology*, **93**, 568–75.

Hampton, R.Y., Holz, R.W. & Goldstein, I.J. (1980). Phospholipid, glycolipid, and ion dependencies of concanavalin A- and *Ricinus communis* agglutinin I-induced agglutination of lipid vesicles. *The Journal of Biological Chemistry*, **255**, 6766–71.

Hankins, C.N., Kindinger, J.I. & Shannon, L.M. (1979). Legume lectins. I. Immunological cross-reactions between the enzymic lectin from mung beans and other well characterized legume lectins. *Plant Physiology*, **64**, 104–7.

—(1980a). Legume α-galactosidase forms devoid of hemagglutinin activity. *Plant Physiology*, **66**, 375–8.

—(1980b). Legume α-galactosidases which have hemagglutinin properties. *Plant Physiology*, **65**, 618–22.

—(1987). The lectins of *Sophora japonica*. I. Purification, properties and *N*-terminal amino acid sequences of two lectins from leaves. *Plant Physiology*, **83**, 825–9.

—(1988). The lectins of *Sophora japonica*. II. Purification, properties and *N*-terminal amino acid sequences of five lectins from bark. *Plant Physiology*, **86**, 67–70.

Hara, T., Tsukamoto, I. & Miyoshi, M. (1983). Oral toxicity of kintoki bean (*Phaseolus vulgaris*) lectin. *Journal of Nutrition Sciences and Vitaminology*, **29**, 589–99.

Hara, T., Mukunoki, Y., Tsukamoto, I., Miyoshi, M. & Hasegawa, K. (1984). Susceptibility of kintoki bean lectin to digestive enzymes *in vitro* and its behaviour in the digestive organs of mouse *in vivo*. *Journal of Nutrition Science and Vitaminology*, **30**, 381–94.

Harel, W., Banay, Y. & Nelken, D. (1981). Skin allograft survival in lentil lectin-treated mice and rats. *Transplantation*, **32**, 69–71.

Harley, S.M. & Beevers, L. (1988). Terminal *N*-acetylglucosamine containing proteins and *N*-glucosamine binding proteins from organelles and membranes of developing pea (*Pisum sativum* L.) cotyledons. *The Journal of Plant Physiology*, **133**, 629–34.

Harley, S.M. & Lord, J.M. (1985). *In vitro* endoproteolytic cleavage of castor bean lectin precursors. *Plant Science*, **41**, 111–16.

Harrison, P.J., Hultborn, H., Jankowska, E., Katz, R., Storai, B. & Zytnicki, D. (1984). Labelling of interneurons by retrograde transsynaptic transport of horse-radish peroxidase from motoneurons in rats and cats. *Neuroscience Letters*, **45**, 15–19.

Haya, S.K. (1988). Transneuronal transport of WGA–HRP in immature rat visual pathways. *Developmental Brain Research*, **38**, 83–8.

Hedo, J.A., Harrison, L.C. & Roth, J. (1981). Binding of insulin receptors to lectins: evidence for common carbohydrate determinants on several membrane receptors. *Biochemistry*, **20**, 3385–93.

Helm, R.M. & Froese, A. (1981) Binding of the receptors for IgE by various lectins. *International Archives of Allergy and Applied Immunology*, **65**, 81–4.

Hemperley, J.J. & Cunningham, B.A. (1983). Circular permutation of amino acid sequences among legume lectins. *Trends in Biochemical Sciences*, **5**, 100–2.

Hendriks, H.G., Koninkx, J.F., Draaijer, M., van Dijk, J.E., Raaijmakers, J.A. & Mouwen, J.M. (1987). Quantitative determination of the lectin binding capacity of small intestinal brush border membrane. An enzyme linked sorbent assay (ELLSA). *Biochimica et Biophysica Acta*, **905**, 371–5.

Henis, Y.I. & Elson, E.L. (1981). Inhibition of the mobility of mouse lymphocyte surface immunoglobulins by locally bound concanavalin A. *Proceedings of the National Academy of Sciences, USA*, **78**, 1072–6.

Herman, E.M. & Shannon, L.M. (1984a). Immunocytochemical localization of concanavalin A in developing jack-bean cotyledons. *Planta*, **161**, 97–104.

—(1984b). The role of the Golgi apparatus in the deposition of the seed lectin of *Bauhinia purpurea* (Leguminosae). *Protoplasma*, **121**, 163–70.

—(1985). Accumulation and subcellular localization of α-galactosidase-hemagglutinin in developing soybean cotyledons. *Plant Physiology*, **77**, 886–90.

Herman, E.M., Hankins, C.N. & Shannon, L.M. (1988). Bark and leaf lectins from *Sophora japonica* are sequestered in protein-storage vacuoles. *Plant Physiology*, **86**, 1027–31.

Heslop-Harrison, J. (1978). Recognition and response in the pollen-stigma interaction. *Symposium of the Society for Experimental Biology*, **32**, 121–38.

Hietanen, J. & Salo, O.P. (1984). Binding of four lectins to normal human oral mucosa. *Scandinavian Journal of Dental Research*, **92**, 443–7.

Higgins, T.J.V., Chandler, P.M., Zurawski, G., Button, S.C. & Spencer, D. (1983a). The biosynthesis and primary structure of pea seed lectin. *The Journal of Biological Chemistry*, **258**, 9544–9.

Higgins, T.J.V., Chrispeels, M.J., Chandler, P.M. & Spencer, D. (1983b). Intracellular sites of synthesis and processing of lectin in developing pea cotyledons. *The Journal of Biological Chemistry*, **258**, 9550–3.

Higuchi, M., Suga, M. & Iwai, K. (1983) Participation of lectin in biological effects of raw winged bean seeds on rats. *Agricultural and Biological Chemistry*, **47**, 1979–86.

Higuchi, M., Tsuchiya, I. & Iwai, K. (1984). Growth inhibition and small intestinal lesions in rats after feeding with isolated wing bean lectin. *Agricultural and Biological Chemistry*, **48**, 695–701.

Hilgert, I., Horejsi, V., Angelisova, P. & Kristofova, H. (1980). Lentil lectin effectively induces allotransplantation tolerance in mice. *Nature*, London, **284**, 273–5.

Hoekstra, D. & Duzgunes, N. (1986). *Ricinum communis* agglutinin-mediated agglutination and fusion of glycolipid-containing phospholipid vesicles: Effect of carbohydrate head group size, calcium ions and spermine. *Biochemistry*, **25**, 1321–30.

Hoffman, L.M. & Donaldson, D.D. (1985). Characterization of two *Phaseolus vulgaris* phytohemagglutinin genes closely linked on the chromosome. *EMBO Journal*, **4**, 883–9.

Hoffman, L.M., Ma, Y. & Barker, R.F. (1982). Molecular cloning of *Phaseolus vulgaris* lectin mRNA and use of cDNA as a probe to estimate lectin transcription levels in various tissues. *Nucleic Acids Research*, **10**, 7819–28.

Hook, W.A., Dougherty, S.F. & Oppenheim, J.J. (1974). Release of histamine from hamster mast cells by concanavalin A and phytohemagglutinin. *Infection and Immunity*, **9**, 903–8.

Horejsi, V. & Kocourek, J. (1974). Studies on phytohemagglutinins. XVII. Some properties of the anti-H-phytohemagglutinin of the furze seeds (*Ulex europeaus* L.). *Biochimica et Biophysica Acta*, **336**, 329–37.

Horejsi, V., Haskovec, C. & Kocourek, J. (1978). Studies on lectins. XXXVIII. Isolation and characterization of the lectin from the locust bark (*Robinia pseudoacacia* L.). *Biochimica et Biophysica Acta*, **532**, 98–104.

Horigome, K., Tamori-Natori, Y., Inoue, K. & Nogima, S. (1986). Effect of serine phospholipid structure on the enhancement of concanavalin A-induced degranulation in rat mast cells. *The Journal of Biochemistry*, **100**, 571–9.

Horisberger, M. & Vonlanthen, M. (1980). Ultrastructural localization of soybean agglutinin on thin sections of *Glycine max* (soybean) var. Altona by the gold method. *Histochemistry*, **65**, 181–6.

Hosselet, M., Van Driessche, E., Van Poucke, M. & Kanarek, L. (1983). Purification and characterization of an endogenous root lectin from *Pisum sativum* L. In *Lectins, Biology, Biochemistry, Clinical Biochemistry* (Bog-Hansen, T.C. & Spengler, G.A., eds), vol. 3, pp. 549–58. Walter de Gruyter, Berlin and New York.

Howard, I.K., Sage, H.J. & Horton, C.S. (1972). Studies on the appearance and location of hemagglutinins from a common lentil during the life cycle of the plant. *Archives of Biochemistry and Biophysics*, **149**, 323–6.

Hume, D.A. & Weidemann, M.J. (1980). *Mitogenic Lymphocyte Transformation*. Elsevier/North Holland Biomedical Press, Amsterdam.

Hutchinson, F.J. & Jones, M.N. (1988). Lectin-mediated targeting of liposomes to a model surface. An ELISA method. *FEBS Letters*, **234**, 493–6.

Hwang, D.L., Yang, W.K. & Foard, D. (1978). Rapid release of protease inhibitors from soybeans. Immunochemical quantitation and parallels with lectins. *Plant Physiology*, **61**, 30–4.

Ichev, K. & Chouchkov, C. (1983). Radioautographic study of leucine uptake and transmission of proteins by rat enterocytes following con A binding. *Acta Histochemistry*, **72**, 181–6.

Ichev, K. & Ovtscharoff, W. (1981). Concanavalin A binding sites on the intestinal microvillous membrane of rat. *Acta Histochemistry*, **69**, 119–24.

Irimura, T., Kawaguchi, T., Terao, T. & Osawa, T. (1975). Carbohydrate binding specificity of the so-called galactose-specific phytohemagglutinins. *Carbohydrate Research*, **39**, 317–27.

Ishiguro, M., Mitarai, M., Harade, H., Sekine, I., Nishimori, T. & Kikutani, M. (1983). Biochemical studies on the oral toxicity of ricin. I. Ricin administered orally can impair sugar absorption by the rat small intestine. *Chemical Pharmocology Bulletin*, **31**, 3222–7.

Ishiguro, M., Harada, H., Ichiki, O., Sekine, I., Nishimori, T. & Kikutani, M. (1984). Effects of ricin, a protein toxin, on glucose absorption by the rat small intestine (Biochemical studies on the oral toxicity of ricin II). *Chemical Pharmacology Bulletin*, **32**, 31141–7.

Ishii, S.I., Abe, Y., Tanaka, I. & Saito, M. (1984). Alteration of quaternary structure and biological activity of concanavalin A. *Journal of Protein Chemistry*, **3**, 63–71.

Ishizuka, T., Ito, Y., Mori, K., Nagao, S., Nagata, K. & Nozawa, Y. (1988). Concanavalin A or phorbol ester-induced translocation of protein kinase C in thymoma cells from a patient with myasthenia gravis. *Clinica Chimica Acta*, **172**, 109–16.

Itaya, S.K. (1988). Transneuronal transport of WGA–HRP in immature rat visual pathways. *Developmental Brain Research*, **38**, 83–8.

Ito, Y. (1986). Occurrence of lectins in leaves and flowers of *Sophora japonica*. *Plant Science*, **47**, 77–82.

Iyer, P.N.S., Wilkinson, K.D. & Goldstein, I.J. (1976). An *N*-acetyl-D-glucosamine binding lectin from *Bandeiraea simplicifolia* seeds. *Archives of Biochemistry and Biophysics*, **177**, 330–3.

Jaffé, W.G. (1980). Hemagglutinins (lectins). In *Toxic Constituents of Plant Foodstuffs*, 2nd Edition (Liener, I.E. ed.), pp 73–102. Academic Press, New York.

Jaffé, W.G., Levy, A. & Gonzales, I.D. (1974). Isolation and partial characterization of bean phytohemagglutinins. *Phytochemistry*, **13**, 2685–93.

Janzen, D.H. (1981). Lectins and plant-herbivore interactions. *Recent Advances in Phytochemistry*, **15**, 241–58.

Janzen, D.H., Juster, H.B. & Liener, I.R. (1976). Insecticidal action of the phytohemagglutinin in black beans on a Bruchid beetle. *Science*, **192**, 795–6.

Janzen, D.H., Ryan, C.A., Liener, I.E. & Peace, G. (1986). Potentially defensive proteins in mature seeds of 59 species of tropical Leguminosae. *Journal of Chemical Biology*, **12**, 1469–80.

Jayne-Williams, D.J. & Hewitt, D. (1972). The relationship between intestinal microflora and the effects of diets containing raw navy bean (*Phaseolus vulgaris*) on the growth of Japanese quail (*Coturnix coturnix japonica*). *Journal of Applied Bacteriology*, **35**, 331–44.

Jayne-Williams, D.J. & Burgess, C.D. (1974). Further observations on the toxicity of navy bean (*Phaseolus vulgaris*) for Japanese quail (*Coturnix coturnix japonica*). *Journal of Applied Bateriology*, **37**, 149–69.

Jermyn, M.A. & Yeow, Y.M. (1975). A class of lectins present in the tissues of seed plants. *Australian Journal of Plant Physiology*, **2**, 501–31.

Jindal, S., Soni, G.L. & Singh, R. (1982). Effect of feeding of lectins from lentils and peas on the intestinal and hepatic enzymes of albino rats. *Journal of Plant Foods*, **4**, 95–103.

—(1984). Biochemical and histopathological studies in albino rats fed on soyabean lectin. *Nutrition Reports International*, **29**, 95–106.

Johns, C.O. & Finks, A.J. (1920*a*). The deficiency of cysteine in proteins of the genus *Phaseolus. Science*, **52**, 414.

—(1920*b*). Studies in nutrition. II. The role of cystine in nutrition as exemplified by nutrition experiments with the proteins of navy bean, *Phaseolus vulgaris. The Journal of Biological Chemistry*, **41**, 379–89.

Johnson, B.A. & Rigas, D.A. (1972). Tryptic glycopeptides of a mitogenic erythro- and lymphoagglutinating phytohemagglutinin (PPHA) from red kidney beans (*Phaseolus vulgaris* L.). *Physiological Chemistry and Physics*, **4**, 245–56.

Johnson, L.R., Tseng, C.C., Tipnis, U.R. & Haddox, M.K. (1988). Gastric mucosal ornithine decarboxylase: localization and stimulation by gastrin. *The American Journal of Physiology*, **255**, G304–12.

Johnson, L.R., Tseng, C.C., Wang, P., Tipnis, U.R. & Hadox, M.K. (1989). Mucosal ornithine decarboxylase in the small intestine: localization and stimulation. *The American Journal of Physiology*, **256**, G624–30.

Kaibuchi, K., Takai, Y. & Nishizuka, Y. (1985). Protein kinase C and calcium ion in mitogenic response of macrophage-depleted human peripheral lymphocytes. *The Journal of Biological Chemistry*, **260**, 1366–9.

Kakaiya, R.M., Kiraly, T.L. & Cable, R.G. (1988). Concanavalin A induces patching/capping of the platelet membrane glycoprotein IIb/IIIa complex. *Thrombosis and Haemostasis*, **59**, 281–3.

Kakinuma, Y., Sakamaki, Y., Ito, K., Cragoe, E.J. & Igarashi, K. (1987). Relationship among activation of the Na^+/H^+ antiporter, ornithine decarboxylase induction, and DNA synthesis. *Archives of Biochemistry and Biophysics*, **259**, 171–8.

Kaku, H., Peumans, W.J. & Goldstein, I.J. (1990). Isolation and characterisation of a second lectin (SNA-II) present in Elderberry (*Sambucus nigra* L) bark. *Archives of Biochemistry and Biophysics*, **237**, 255–62.

Kaplowitz, P.B. (1985). Wheat germ agglutinin and concanavalin A inhibit the response of human fibroblasts to peptide growth factors by a post-receptor mechanism. *The Journal of Cellular Physiology*, **124**, 474–80.

Kaplowitz, P.B. & Haar, J.L. (1988). Antimitogenic actions of lectins in cultured human fibroblasts. *The Journal of Cellular Physiology*, **136**, 13–22.

Kato, G., Maruyama, Y. & Nakamura, M. (1981). Involvement of lectins in *Rhizobium*-pea recognition. *Plant Cell Physiology*, **22**, 759–72.

Katsumoto, T. & Kurimura, T. (1988). Ultrastructural localization of concanavalin A receptors in the plasma membrane: association with underlaying actin filaments. *Biology of the Cell*, **62**, 1–10.

Katzen, H.M., Vicario, P.P., Mumford, R.A. & Green, B.G. (1981). Evidence that the insulin-like activities of concanavalin A and insulin are mediated by a common insulin receptor linked effector system. *Biochemistry*, **20**, 580–9.

Kauss, H. & Glaser, C. (1974). Carbohydrate-binding proteins from plant cell walls and their possible involvement in extension growth. *FEBS Letters*, **45**, 304–7.

Kawakami, K., Yamamoto, Y. & Onoue, K. (1988). Effect of wheat germ agglutinin on T lymphocyte activation. *Microbiological Immunology*, **32**, 413–22.

Kawano, M., Iwato, K. & Kuramoto, A. (1985). Identification and characterization of a B cell growth inhibitory factor (BIF) on BCGF-dependent B cell proliferation. *The Journal of Immunology*, **134**, 375–81.

Keenan, K.P., Sharpnack, D.D., Collins, H., Formal, S.B. & O'Brien, A.D. (1986). Morphologic evaluation of the effects of Shiga toxin and E. coli Shiga-like toxin on the rabbit intestine. *The American Journal of Pathology*, **125**, 69–80.

Key, B. & Giorgi, P.P. (1986). Selective binding of soybean agglutinin to the olfactory system of *Xenopus*. *Neuroscience*, **18**, 507–15.

Kijne, J.W., Smit, G., Diaz, C.L. & Lugtenberg, B.J.J. (1986). Attachment of *Rhizobium leguminosarum* to pea root hair tips. In: *Recognition in Microbe-Plant Symbiotic and Pathogenic Interactions* (Lugtenberg, B. ed.), pp. 101–11. Springer-Verlag KG, Berlin.

—(1988). Lectin-enhanced accumulation of manganase-limited *Rhizobium leguminosarum* cells on pea root hair tips. *Journal of Bacteriology*, **170**, 2994–3000.

Kilpatrick, D.C. (1980*a*). Purification and some properties of a lectin from the fruit juice of the tomato (*Lycopersicon esculentum*). *The Biochemical Journal*, **185**, 169–72.

—(1980*b*). Isolation of a lectin from the pericarp of potato (*Solanum tuberosum*) fruits. *The Biochemical Journal*, **191**, 273–5.

—(1988). Accessory cell paradox: monocytes enhance or inhibit lectin-mediated human T-cell lymphocyte proliferation depending on the choice of mitogen. *Scandinavian Journal of Immunology*, **28**, 247–9.

Kilpatrick, D.C. & McCurrach, P.W. (1987). Wheat germ agglutinin is mitogenic, non-mitogenic and antimitogenic for human lymphocytes. *Scandinavian Journal of Immunology*, **25**, 343–8.

Kilpatrick, D.C., Yeoman, M.M. & Gould, A.R. (1979). Tissue and subcellular distribution of the lectin from *Datura stramonium* (Thorn apple). *The Biochemical Journal*, **184**, 215–19.

Kilpatrick, D.C., Graham, C. & Urbaniak, S.J. (1986). Inhibition of human lymphocyte transformation by tomato lectin. *Scandinavian Journal of Immunology*, **24**, 11–19.

Kilpatrick, D.C., Pusztai, A., Grant, G., Graham, C. & Ewen, S.W.B. (1985). Tomato lectin resists digestion in the mammalian alimentary canal and binds to intestinal villi without deleterious effects. *FEBS Letters*, **185**, 299–305.

Kimura, A. & Ersson, B. (1981). Activation of T lymphocytes by lectins and carbohydrate-oxidising reagents viewed as an immunological recognition of cell-surface modifications seen in the context of 'self' major histocompatibility complex antigens. *European Journal of Immunology*, **11**, 475–83.

Kimura, T., Nakata, S., Harada, Y. & Yoshida, A. (1986). Effect of ingested winged bean lectin on gastrointestinal function in the rat. *Journal of Nutrition and Science of Vitaminology*, **32**, 101–10.

King, A.C. & Cuetracasas, P. (1981). Peptide hormone-induced receptor mobility, aggregation and internalization. *New England Journal of Medicine*, **305**, 77–88.

King, T.P., Pusztai, A. & Clarke, E.M.W. (1980*a*). Immunocytochemical localization of ingested kidney bean (*Phaseolus vulgaris*) lectins in rat gut. *The Histochemical Journal*, **12**, 201–8.

—(1980*b*). Kidney bean (*Phaseolus vulgaris*) lectin-induced lesions in rat small intestine. 1. Light microscope studies. *The Journal of Comparative Pathology*, **90**, 585–95.

—(1982). Kidney bean (*Phaseolus vulgaris*) lectin-induced lesions in rat small intestine. 2. Ultrastructural studies. *The Journal of Comparative Pathology*, **92**, 357–73.

King, T.P., Begbie, R. & Cadenhead, A. (1983). Nutritional toxicity of raw kidney bean in pigs. Immunocytochemical and cytopathological studies on the gut and the pancreas. *The Journal of Science of Food and Agriculture*, **34**, 1404–12.

King, T.P., Pusztai, A., Grant, G. & Slater, D. (1986). Immunogold localization of ingested kidney bean (*Phaseolus vulgaris*) lectins in epithelial cells of the rat small intestine. *The Histochemical Journal*, **18**, 413–20.

Kinoshita, S., Yoshii, K. & Tonegawa, Y. (1988). Specific binding of lectins with the nucleus of the sea urchin embryo and changes in lectin affinity of the embryonic chromatin during the course of development. *Experimental Cell Research*, **175**, 148–57.

Klajic, Z., Schroder, H.C., Rottmann, M., Cuperlovic, M., Movsesian, M., Uhlenbruck, G., Gasic, M., Zahn, R.K. & Muller, W.E.G. (1987). A D-mannose-specific lectin from *Gerardia savaglia* that inhibits nucleocytoplasmic transport of mRNA. *European Journal of Biochemistry*, **69**, 97–104.

Kleinman, R.E. & Walker, W.A. (1984). Antigen processing and uptake from the intestinal tract. *Clinical Reviews on Allergy*, **2**, 25–37.

Knox, R.B., Clarke, A., Harrison, S., Smith, P. & Marchalonis, J.J. (1976). Cell recognition in plants: Determinants of the stigma surface and their pollen interactions. *Proceedings of the National Academy of Sciences, USA*, **73**, 2788–92.

Kocourek, J. & Horejši, V. (1983). Note on the recent discussion on definition of the term 'lectin'. In: *Lectins, Biology, Biochemistry, Clinical Biochemistry* (Bog-Hansen, T.C. & Spengler, G.A., eds), pp. 3–6. Walter de Gruyter, Berlin and New York.

Koenig, S. & Hoffman, M.K. (1979). Bacterial lipopolysaccharide activates suppressor B lymphocytes. *Proceedings of the National Academy of Sciences, USA*, **76**, 4608–12.

Kolberg, J. & Sollid, L. (1985). Lectin activity of gluten identified as wheat germ agglutinin. *Biochemical and Biophysical Research Communications*, **130**, 867–72.

Kornfeld, R., Gregory, W.T. & Kornfeld, S. (1972). Red kidney bean (*Phaseolus vulgaris*) phytohemagglutinin. In: *Methods in Enzymology* (Ginzburg, V. ed.), 28, part B, pp. 344–9. Academic Press, New York.

Köttgen, E., Volk, B., Kluge, F. & Gerok, W. (1982). Gluten, a lectin with oligomannosyl specificity and the causative agent of gluten-sensitive enteropathy. *Biochemical and Biophysical Research Communications*, **130**, 867–72.

Köttgen, F., Kluge, F., Volk, B. & Gerok, W. (1983). The lectin properties of gluten as the basis of pathomechanism of gluten-sensitive enteropathy. *Klinische Wochenschrift*, **61**, 111–12.

Kouchalakos, R.N., Bates, O.J., Bradshaw, R.A. & Hapner, K.D. (1984). Lectin from sainfoin (*Onobrychis viciifolia* scop.). Complete amino acid sequence. *Biochemistry*, **23**, 1824–30.

Kronis, K.A. & Carver, J.P. (1985). Wheat germ agglutinin dimers bind sialyloligosaccharides at four sites in solution: proton nuclear magnetic resonance temperature studies at 360 MHz. *Biochemistry*, **24**, 826–33.

Kronke, M., Leonard, W.J., Kepper, J.M. & Greene, W.C. (1985). Sequential expression of genes involved in human T lymphocyte growth and differentiation. *The Journal of Experimental Medicine*, **162**, 1693–8.

Kumagai, N., Benedict, S.H., Mills, G.B. & Gelfand, E.W. (1988). Comparison of phorbol ester/calcium ionophore and phytohemagglutinin-induced signalling in human T lymphocytes. Demonstration of interleukin 2-independent transferrin receptor gene expression. *The Journal of Immunology*, **140**, 37–43.

Kurisu, M., Yamazaki, M. & Mizuno, D. (1980). Induction of macrophage-mediated tumor lysis by the lectin wheat germ agglutinin. *Cancer Research*, **40**, 3798–803.

Kyte, J. & Doolittle, R.F. (1982). A simple method for displaying the hydropathic character of a protein. *Journal of Molecular Biology*, **157**, 105–32.

Lafont, J., Rouanet, J.M., Gabrion, J., Assouad, J.L., Zambonino Infante, J.L. & Resancon, P. (1988). Duodenal toxicity of dietary *Phaseolus vulgaris* lectins in the rat: An integrative assay. *Digestion*, **41**, 83–93.

Lamb, F.I., Roberts, L.M. & Lord, J.M. (1985). Nucleotide sequence of cloned cDNA coding for preproricin. *European Journal of Biochemistry*, **148**, 265–70.

Landsteiner, K. (1962). *The Specificity of Serological Reactions*. Revised edition, Dover, New York.

Landsteiner, K. & Raubitschek, H. (1908). Reobachtungen uber Hamolyse und Hamagglutination. *Zentralblatt für Bakteriologie und Parasitenkunde, Infektionskrankheiten und Hygiene. Abteilung. 1. Originale*, **45**, 660–7.

Lang, E., Kohl, U., Schirrmacher, V., Brossmer, R. & Altevogt, P. (1987). Structural basis for altered soybean agglutinin lectin binding between a murine metastatic lymphoma and an adhesive low malignant variant. *Experimental Cell Research*, **173**, 232–43.

Langone, J.J. & Ejzemberg, R. (1981). Succinylated and acetylated concanavalin A activate the classical complement pathway. *Biochemical and Biophysical Research Communications*, **99**, 768–74.

Law, I.J. & Strijdom, B.W. (1984a). Properties of lectins in the root and seed of *Lotononis bainesii*. *Plant Physiology*, **74**, 773–8.

—(1984b). Role of lectins in the specific recognition of *Rhizobium* by *Lotononis bainesii*. *Plant Physiology*, **74**, 779–85.

Lea, T., Smeland, E., Funderud, S., Vartdal, F., Davies, C., Beiske, K. & Ugelstad, J. (1986). Characterization of human mononuclear cells after positive selection with immunomagnetic particles. *Scandinavian Journal of Immunology*, **23**, 509–19.

Leach, J.E., Cantrell, M.A. & Sequeira, L. (1982). Hydroxyproline-rich bacterial agglutinin from the potato. Extraction, purification and characterization. *Plant Physiology*, **70**, 1353–8.

Leavitt, R.D., Felsted, R.L. & Bachur, N.R. (1977). Biological and biochemical properties of *Phaseolus vulgaris* isolectins. *The Journal of Biological Chemistry*, **252**, 2961–6.

Lee, C.L., McFarland, D.J. & Wolpaw, J.R. (1988). Retrograde transport of the lectin *Phaseolus vulgaris* leucoagglutinin (PHA-L) by rat spinal motoneurons. *Neuroscience Letters*, **86**, 133–8.

Leivestadt, T., Halvorsen, R., Gaudernack, G. & Thorsby, E. (1988). Requirements for phytohemagglutinin activation of resting pure CD4+ and CD8+ T cells. *Scandinavian Journal of Immunology*, **27**, 565–72.

Lev, B., Ward, H., Keusch, G.T. & Pereira, M.E.A. (1986). Lectin activation in *Giardia lamblia* by host protease: a novel host–parasite interaction. *Science*, **232**, 71–3.

LeVine, D., Kaplan, M.J. & Greenaway, P.J. (1972). The purification and characterization of wheat germ agglutinin. *The Biochemical Journal*, **129**, 847–56.

Liener, I.E. (1986). Nutritional significance of lectins in the diet. In *The Lectins* (Liener, I.E., Sharon, N. & Goldstein, I.J. eds), pp. 527–52. Academic Press, Orlando.

Lima, M.S., Oliveira, M.C., Moreira, R.A. & Prouvost-Danon, A. (1983). Metabolic energy-dependent exocytosis of mouse mast cells by lectin of *Dioclea grandiflora* (Mart). *Biochemical Pharmacology*, **34**, 4169–70.

Ling, E.A. & Leong, S.K. (1987). Effects of intraneuronal injection of *Ricinus communis* agglutinin-60 into the rat vagus nerve. *Journal of Neurocytology*, **16**, 373–87.

Lipkowitz, S., Greene, W.C., Rubin, A.L., Novogrodsky, K. & Stenzel, K.H. (1984). Expression of receptors for interleukin 2: Role in the commitment of T lymphocytes to proliferate. *The Journal of Immunology*, **132**, 31–7.

Lis, H. & Sharon, N. (1981). Lectins in higher plants. In *The Biochemistry of Plants* (Marucs, A. ed.), vol.6, pp. 371–447. Academic Press, New York.

—(1986). Biological properties of lectins. In *Lectins* (Liener, J.E., Sharon, N. & Goldstein, J.J. eds), pp. 265–91. Academic Press, Orlando.

Lis, H., Sharon, N. & Katchalski, E. (1964). Isolation of a mannose-containing glycopeptide from soybean hemagglutinin. *Biochimica et Biophysica Acta*, **83**, 376–8.

Lis, H., Fridman, C., Sharon, N. & Katchalski, E. (1966). Multiple hemagglutinins in soybean. *Archives of Biochemistry and Biophysics*, **117**, 301–9.

Livingstone, J.N. & Purvis, B.J. (1980). Effects of wheat germ agglutinin on insulin binding and insulin sensitivity of fat cells. *The American Journal of Physiology*, **238**, E267–75.

Lonngren, J., Goldstein, I.J. & Bywater, R. (1976). Cross-linked guaran: A versatile immunosorbent for D-galactopyranosyl binding lectins. *FEBS Letters*, **63**, 31–4.

Lord, J.M. (1985a). Synthesis and intracellular transport of lectin and storage protein precursors in endosperm from castor bean. *European Journal of Biochemistry*, **146**, 403–9.

—(1985b). Precursors of ricin and *Ricinus communis* agglutinin. Glycosylation and processing during synthesis and intracellular transport. *European Journal of Biochemistry*, **146**, 411–16.

Lord, J.M. (1987). The use of cytotoxic plant lectins in cancer therapy. *Plant Physiology*, **85**, 1–3.

Lord, J.M., Roberts, L.M., Thorpe, P.R. & Vitetta, E.S. (1985). Immunotoxins. *Trends in Biotechnology*, **3**, 175–9.

Lorenz-Meyer, H., Roth, P., Elsasser, P. & Hahn, U. (1985). Cytotoxicity of lectins on rat intestinal mucosa enhanced by neuramidase. *European Journal of Clinical Investigations*, **15**, 227–34.

Lorenzsonn, V. & Olsen, W.A. (1982). *In vivo* responses of rat intestinal epithelium to intraluminal dietary lectins. *Gastroenterology*, **82**, 838–48.

Lotan, R., Lis, H., Rosenwasser, A., Novogrodsky, A. & Sharon, N. (1973). Enhancement of the biological activities of soybean agglutinin by cross-linking with glutaraldehyde. *Biochemical and Biophysical Research Communications*, **55**, 1347–55.

Lotan, R., Siegelman, H.W., Lis, H. & Sharon, N. (1974). Subunit structure of soybean agglutinin. *The Journal of Biological Chemistry*, **249**, 121–24.

Lotan, R., Sharon, N. & Mirelman, D. (1975a). Interaction of wheat germ agglutinin with bacterial cells and cell wall polymers. *European Journal of Biochemistry*, **55**, 252–62.

Lotan, R., Cacan, R., Cacan, M., Debray, H., Carter, W.G. & Sharon, N. (1975b). On the presence of two types of subunit in soybean agglutinin. *FEBS Letters*, **57**, 100–3.

Ludwin, D. & Singal, D.P. (1986). Prolonged allograft survival resulting from donor pretreatment with phytohemagglutinin-P. *Transplantation*, **41**, 120–2.

Luk, G.D. & Baylin, S.B. (1983). Polyamines and intestinal growth – increased polyamine biosynthesis after jejunectomy. *The American Journal of Physiology*, **245**, G656–60.

Luk, G.D. & Yang, P. (1987). Polyamines in intestinal and pancreatic adaptation. *Gut*, **28**, S1 95–101.

—(1988). Distribution of polyamines and their biosynthetic enzymes in intestinal adaptation. *The American Journal of Physiology*, **254**, G194–200.

Luk, G.D., Marton, L.J. & Baylin, S.B. (1980). Ornithine decarboxylase is important in intestinal mucosal maturation and recovery from injury in rats. *Science*, **210**, 195–8.

Luning, O. & Bartels, W. (1926). The toxicity of white beans (*Phaseolus vulgaris*). *Zeitschrift für Untersuchung der Lebensmittel*, **51**, 220–8.

Lutsik, M.D. & Antonyuk, V.A. (1982). New fucose-specific lectin from the bark of *Laburnum anagyroides* Medik: purification, properties and immunochemical specificity. *Biochemistry (USSR)*, **47**, 1448–53.

Lynn, K.R. & Clevette-Radford, N.A. (1986). Lectins from latices of Euphorbia and Elaeophorbia species. *Phytochemistry*, **25**, 1553–7.

Majerus, P.W., Connolly, T.M., Deckmyn, H., Ross, T.S., Bross, T.E., Ishii, H., Bansal, V.S. & Wilson, D.B. (1986). The metabolism of phosphoinositide-derived messenger molecules. *Science*, **234**, 1519–26.

Maliarik, M.J. & Goldstein, I.J. (1988). Photoaffinity labeling of the adenine binding site of the lectins from lima bean, *Phaseolus lunatus,* and the kidney bean, *Phaseolus vulgaris*. *The Journal of Biological Chemistry*, **263**, 11274–9.

Maliarik, M., Plessas, N.R., Goldstein, I.J., Musci, G. & Berliner, L.J. (1989). ESR and fluorescence studies on the adenine binding site of lectins using spin-labeled analogue. *Biochemistry*, **28**, 912–17.

Manage, L., Joshi, A. & Sohonie, K. (1972). Toxicity to rats and mice of purified phytohaemagglutinins from four Indian legumes. *Toxicon*, **10**, 89–91.

Manen, J.F. & Pusztai, A. (1982). Immunocytochemical localization of lectins in cells of *Phaseolus vulgaris* L. seeds. *Planta*, **155**, 328–34.

Mansfield, M.A., Peumans, W.J. & Raikhel, N.V. (1988). Wheat germ agglutinin is synthesized as a glycosylated precursor. *Planta*, **173**, 482–9.

Marcus, S.E., Maycox, P.R. & Bowles, D.J. (1989). Con A-binding polypeptides in jackbean cotyledons. *Phytochemistry*, **28**, 333–6.

Margolis, T.P. & LaVail, J.H. (1984). Further evidence in support the selective uptake and anterograde transport of [^{125}I] wheat germ agglutinin by chick retinal ganglion cells. *Brain Research*, **324**, 21–7.

Marinkovich, V.A. (1964). Purification and characterization of the hemagglutinin present in potatoes. *Journal of Immunology*, **93**, 732–41.

Marjanovic, S., Wielburski, A. & Nelson, B.D. (1988). Effect of phorbol myristate acetate and concanavalin A on the glycolytic enzymes of human peripheral lymphocytes. *Biochimica et Biophysica Acta*, **970**, 1–6.

Markley, K., Thornton, S., Smallman, E. & Markley, P. (1969). Prolongation of skin graft survival and induction of resistance to infection by phytohemagglutinin. *Transplantation*, **8**, 258–64.

Matsui-Yuasa, I., Otani, S. & Morisawa, S. (1987). Role of protein kinase C in phytohemagglutinin-stimulated induction of spermidine/spermine N^1-acetyltransferase. *Biochemistry International*, 15, 997–1003.

Mazurier, J., Legrand, D., Hu, W.L., Montreuil, J. & Spik, G. (1989). Expression of human lactotransferrin receptors in phytohemagglutinin-stimulated human peripheral blood lymphocytes. Isolation of the receptors by antiligand-affinity chromatography. *European Journal of Biochemistry*, 187, 481–7.

Mäkelä, D. (1957). Studies in hemagglutinins of leguminosae seeds. *Annales Medicinae Experimentalis et Biologiae Fenniae*, 35, Suppl. 11, pp. 1–56.

McCurrach, P.M. & Kilpatrick, D.C. (1988). Dature lectin is both an antimitogen and a co-mitogen acting synergistically with phorbol ester. *Scandinavian Journal of Immunology*, 27, 31–4.

McDonel, J.L. (1980). Binding of *Clostridium perfringens* (^{125}I)enterotoxin to rabbit intestinal cells. *Biochemistry*, 19, 4801–7.

McPherson, A. & Hoover, S. (1979). Purification of mitogenic proteins from Hura crepitans and Robinia pseudoacacia. *Biochemical and Biophysical Research Communications*, 89, 713–20.

Messina, J.L., Hamlin, J. & Larner, J. (1987*a*). Insulin-mimetic actions of wheat germ agglutinin and concanavalin A on specific mRNA levels. *Archives of Biochemistry and Biophysics*, 254, 110–15.

—(1987*b*). Positive interaction between insulin and phorbol esters on the regulation of specific messenger ribonucleic acid in rat hepatoma cells. *Endocrinology*, 121, 1227–32.

Metcalf, T.N. III., Wang, J.L., Schubert, K.R. & Schjindler, M. (1983). Lectin receptors on the plasma membrane of soybean cells. Binding and lateral diffusion of lectins. *Biochemistry*, 22, 3969–75.

Meuer, S.C., Hussey, R.E., Penta, A.C., Fitzgerald, K.A., Stadler, B.M., Schlossman, S.F. & Reinherz, E.L. (1982). Cellular origin of interleukin-2 (IL-2) in man: evidence for stimulus-restricted IL-2 production by T4+ and T8+ T lymphocytes. *Journal of Immunology*, 129, 1076–9.

Mezitis, S.G.E., Stieber, A. & Gonatas, N.K. (1987). Quantitative ultrastructural autoradiographic evidence for the magnitude and early involvement of the Golgi apparatus complex in the endocytosis of wheat germ agglutinin by cultured neuroblastoma. *Journal of Cellular Physiology*, 132, 401–14.

Michell, R.H. (1989). Post-receptor signalling pathways. *Lancet*, i, 765–8.

Miller, J.B., Noyes, C., Heinrikson, R., Kingdon, H.S. & Yachnin, S. (1973). Phytochemagglutinin mitogenic proteins. Structural evidence for a family of isomitogenic proteins. *The Journal of Experimental Medicine*, 138, 939–51.

Miller, J.B., Hsu, R., Heinrikson, R. & Yachnin, S. (1975). Extensive homology between the subunits of the phytohemagglutinin mitogenic proteins derived from *Phaseolus vulgaris*. *Proceedings of the National Academy of Sciences, USA*, 72, 1388–91.

Miller, K. (1983). The stimulation of human B and T lymphocytes by various lectins. *Immunobiology*, 165, 132–46.

Miller, R.C. & Bowles, D.J. (1982). A comparative study of the location of wheat germ agglutinin and its potential receptors in wheat grains. *The Biochemical Journal*, 206, 571–6.

Miltenburg, A.M.M., Meijer-Paape, M.E., Daka, M.R. & Paul, L.C. (1987). Inhibition of T cell cytolytic potential by concanavalin A: a result of activation. *Scandinavian Journal of Immunology*, 26, 555–61.

Mire, A.R., Wickremasinghe, R.G. & Hoffbrand, A.V. (1986a). Phytohemagg-lutinin treatment of T lymphocytes stimulates rapid increases in activity of both particulate and cytostolic protein kinase C. *Biochemical and Biophysical Research Communications*, **137**, 128–43.

—(1986b). Mitogen treatment of permeabilized human T lymphocytes stimulates rapid tyrosine and serine phosphorylation of a 42 kDa protein. *FEBS Letters*, **206**, 53–8.

Mirelman, D., Galun, E., Sharon, N. & Lotan, R. (1975). Inhibition of fungal growth by wheat germ agglutinin. *Nature*, London, **256**, 414–16.

Mishkind, M., Keegstra, K. & Palevitz, B.A. (1980). Distribution of wheat germ agglutinin in young plants. *Plant Physiology*, **66**, 950–5.

Mishkind, M., Raikhel, N.V., Palevitz, B.A. & Keegstra, K. (1982). Immuno-cytochemical localization of wheat germ agglutinin in wheat. *The Journal of Cell Biology*, **92**, 753–64.

Mishkind, M., Palevitz, B.A., Raikhel, N.V. & Keegstra, K. (1983). Localiz-ation of wheat germ agglutinin-like lectins in various species of Grami-neae, *Science*, **220**, 1290–2.

Mizushima, Y., Kosaka, H., Sakuma, S., Kanda, K., Itoh, K., Osugi, T., Mitzushima, A., Hamaoka, T., Yoshida, H., Sobue, K. & Fujiwara, H. (1987). Cyclosporin A inhibits late steps of T lymphocyte activation after transmembrane signalling. *Journal of Biochemistry*, **102**, 1193–201.

Montfort, W., Villafranca, J.E., Monzingo, A.F., Ernst, S.R., Katzin, B., Rutenber, E., Xuong, N.H., Hamlin, R. & Robertus, J.D. (1987). The three-dimensional structure of ricin at 2.8 A. *The Journal of Biological Chemistry*, **262**, 5398–403.

Moraru, I.I., Manciulea, M., Calugaru, A., Ghyka, G. & Popescu, L.M. (1987). Anti-phospholipase C antibodies inhibit the lectin-induced proliferation of human lymphocytes. *Bioscience Reports*, **7**, 731–6.

Moreira, R.A., Barros, A.C.H., Stewart, J.C. & Pusztai, A. (1983). Isolation and characterization of a lectin from the seeds of *Dioclea grandiflora* (Mart.). *Planta*, **158**, 63–9.

Morgan, W.T.J. & Watkins, W.M. (1953). The inhibition of haemagglutinins in plant seeds by human blood group substances and simple sugars. *British Journal of Experimental Pathology*, **34**, 94–103.

Moskaug, J.O., Sandvig, K. & Olsnes, S. (1988). Low pH-induced release of diphtheria toxin A fragment in Vero cells. Biochemical evidence for transfer to the cytosol. *The Journal of Biological Chemistry*, **263**, 2518–25.

Moullier, P., Daveloose, D., Leterrier, F. & Hoebke, J. (1986). Comparative binding of wheat germ agglutinin and its succinylated form on lympho-cytes. *European Journal of Biochemistry*, **161**, 197–204.

Muelenaere, de, H.J.H. (1965). Toxicity and haemagglutinating activity of legumes. *Nature*, London, **206**, 827–8.

Mustelin, T. (1987). GTP dependence of the transduction of mitogenic signals through the T3 complex in T lymphocytes indicates the involvement of a G protein. *FEBS Letters*, **213**, 199–203.

Mustelin, T., Poso, H., Livanainen, A. & Anderson, L.C. (1986). Myo-inositol reverses Li$^+$-induced inhibition of phosphoinositide turnover and ornith-ine decarboxylase induction during early lymphocyte activation. *European Journal of Immunology*, **16**, 859–61.

Nachbar, M.S. & Oppenheim, D.J. (1980). Lectins in the United States diet: a survey of lectins in commonly consumed foods and a review of the literature. *American Journal of Clinical Nutrition*, **33**, 2338–45.

Nachbar, M.S., Oppenheim, J.D. & Thomas, J.O. (1980). Lectins in the United States diet. Isolation and characterization of a lectin from the tomato (*Lycopersicon esculentum*). *The Journal of Biological Chemistry*, **265**, 2056–61.

Nagata, Y. & Burger, M.M. (1972). Wheat germ agglutinin. Isolation and crystallization. *The Journal of Biological Chemistry*, **247**, 2248–50.

—(1974). Wheat germ agglutinin. Molecular characteristics and specificity for sugar binding. *The Journal of Biological Chemistry*, **249**, 3116–22.

Nakata, S. & Kimura, T. (1985). Effect of ingested toxic bean lectins on the gastrointestinal tract in the rat. *The Journal of Nutrition*, **115**, 1621–9.

—(1986). Behaviour of ingested concanavalin A in the gastrointestinal tract of the rat. *Agricultural and Biological Chemistry*, **50**, 645–9.

Neckers, L.M. & Cossman, J. (1983). Transferrin receptor induction in mitogen-stimulated human T lymphocytes is required for DNA synthesis and cell division and is regulated by interleukin 2. *Proceedings of the National Academy of Sciences, USA*, **80**, 3494–8.

Ng, T.B., Li, W.W. & Yeung, H.W. (1989). Effects of lectins with various carbohydrate binding specificities on lipid metabolism in isolated rat and hamster adipocytes. *International Journal of Biochemistry*, **21**, 149–55.

Nicolson, G.L. (1974). Interactions of lectins with animal cell surfaces. *International Review of Cytology*, **39**, 89–190.

Nicolson, G.L., Blaustein, J. & Etzler, M.B. (1974). Characterization of two plant lectins from *Ricinus communis* and their quantitative interaction with murine lymphoma. *Biochemistry*, **13**, 196–204.

Nicolson, G.L., Lacorbiere, M. & Hunter, T.R. (1975). Mechanism of cell entry and toxicity of an affinity-purified lectin from *Ricinus communis* and its differential effects on normal and virus-transformed fibroblasts. *Cancer Research*, **35**, 144–55.

Nishizuka, Y. (1984). The role of protein kinase C in cell surface signal transduction and tumour promotion. *Nature, London*, **308**, 693–8.

Nolte, H., Skov, P.S. & Loft, H. (1987). Pathophysiological role of histamine and clinical aspects. Stimulation of histamine synthesis from tumor cells by concanavalin A and A 23187. *Agents and Actions*, **20**, 291–4.

Novakova, N. & Kocourek, J. (1974). Studies on phytohemagglutinins. XX. Isolation and characterization of hemagglutinins from scarlet runner seeds (*Phaseolus coccineus* L.). *Biochimica et Biophysica Acta*, **359**, 320–33.

Novogrodsky, A. & Ashwell, G. (1977). Lymphocyte mitogenesis induced by mammalian liver protein that specifically binds desialylated glycoproteins. *Proceedings of the National Academy of Sciences, USA*, **74**, 676–8.

Novogrodsky, A. & Katchalski, E. (1973). Transformation of neuraminidase-treated lymphocytes by soybean agglutinin. *Proceedings of the National Academy of Sciences, USA*, **70**, 2515–18.

Novogrodsky, A., Lotan, R., David, A. & Sharon, N. (1975). Peanut agglutinin, a new mitogen that binds to galactosyl sites exposed after neuraminidase treatment. *The Journal of Immunology*, **115**, 1243–8.

Nowell, P. (1960). Phytohemagglutinin: an initiator of mitosis in cultures of normal human leucocytes. *Cancer Research*, **20**, 462–4.

Nsimba-Lubaki, M. & Peumans, W.J. (1986). Seasonal fluctuation of lectins in barks of Elderberry (*Sambucus nigra*) and Black locust (*Robinia pseudoacacia*). *Plant Physiology*, **80**, 747–51.

Nsimba-Lubaki, M., Allen, A.K. & Peumans, W.J. (1986a). Isolation and partial characterization of latex lectins from three species of the genus Euphorbia (Euphorbiceae). *Physiologia Plantarum*, **67**, 193–8.

Nsimba-Lubaki, M., Peumans, W.J. & Allen, A.K. (1986*b*). Isolation and characterization of glycoprotein lectins from the bark of three species of elder, *Sambucus ebulus, S. nigra* and *S. racemosa. Planta*, **168**, 113–18.

Obrig, T.G., Moran, T.P. & Colinas, R.J. (1985). Ribonuclease activity associated with the 60S ribosome-inactivating proteins ricin A, phytolaccin and Shiga toxin. *Biochemical and Biophysical Research Communications*, **130**, 379–84.

Oda, Y. & Minami, K. (1986). Isolation and characterization of a lectin from tulip bulbs *Tulip gesneriana. European Journal of Biochemistry*, **159**, 239–45.

Oda, Y., Minami, K., Ichida, S. & Aonuma, S. (1987). A new agglutinin from the *Tulipa gesneriana* bulbs. *European Journal of Biochemistry*, **165**, 297–302.

Ofek, I. & Sharon, N. (1988). Lectinophagocytosis: a molecular mechanism of recognition between cell surface sugars and lectins in the phagocytosis of bacteria. *Infection and Immunity*, **56**, 539–47.

Olden, K., Parent, J.B. & White, S.L. (1982). Carbohydrate moieties of glycoproteins. A re-evaluation of their function. *Biochimica et Biophysica Acta*, **650**, 209–32.

Oliveira, de, J.T.A. (1986). Seed lectins. The effects of dietary *Phaseolus vulgaris* lectins on the metabolism of monogastric animals. *PhD thesis. University of Aberdeen.*

Oliveira, de, J.T.A., Pusztai, A. & Grant, G. (1988). Changes in organs and tissues induced by feeding of purified kidney bean (*Phaseolus vulgaris*) lectins. *Nutrition Research*, **8**, 943–7.

Olsen, K.W. (1983). Prediction of three-dimensional structure of plant lectins from the domains of concanavalin A. *Biochimica et Biophysica Acta*, **743**, 212–18.

Olsnes, S. & Pihl, A. (1973). Different biological properties of the two constituent peptide chains of ricin, a toxic protein inhibiting protein synthesis. *Biochemistry*, **12**, 3121–6.

—(1977). Abrin, ricin and their associated agglutinins. In *Specificity and Action of Animal and Bacterial and Plant Toxins*, Receptors and Recognition Series B. vol. 1, pp. 171–3. (Cuetracasas, P. ed.). Chapman and Hall, London.

—(1982). Toxic lectins and related proteins. In *The Molecular Actions of Toxins and Viruses* (Cohen, P. & Van Heynigen, S. eds), pp. 51–105. Elsevier Biomedical Press, New York.

Olsnes, S., Refsnes, K. & Pihl, A. (1974). Mechanism of action of the toxic lectins abrin and ricin. *Nature*, London, **249**, 627.

Olsnes, S., Moskaug, J.O., Stenmark, H. & Sandvig, K. (1988). Diphtheria toxin entry: protein translocation in the reverse direction. *Trends in Biochemical Sciences*, **13**, 348–50.

Olson, A.D., Pysher, T.J., Larrosa-Haro, A., Mahmood, A. & Torres-Pinedo, R. (1985). Differential toxicity of RCA$_{II}$ (ricin) on rabbit intestinal epithelium in relation to postnatal maturation. *Pediatric Research*, **19**, 868–72.

Osborn, T.C., Ausloos, K.A., Brown, J.W.S. & Bliss, F.A. (1983). Bean lectins. III. Evidence for greater complexity in the structural model of *Phaseolus vulgaris* lectin. *Plant Science Letters*, **31**, 193–203.

Osborn, T.C., Burrow, M. & Bliss, F.A. (1988). Purification and characterization of Arcelin seed protein from common bean. *Plant Physiology*, **86**, 399–405.

Otani, S., Matsui-Yuasa, I., Hashikawa, K., Kasai, S., Matsui, K. & Morisawa, S. (1986). Synergistic induction of ornithine decarboxylase by diacylglycerol, A 23187 and cholera toxin in guinea pig lymphocytes. *Biochemical and Biophysical Research Communications*, 130, 389–95.

Ottensooser, F. & Silberschmidt, K. (1953). Haemagglutinin anti-N in plant seeds. *Nature, London*, 172, 914.

Ovtscharoff, W. & Ichev, K. (1984). Localization of lectin binding sites on the rat intestinal microvillus membrane. *Acta Histochemistry*, 74, 21–4.

Owens, R.J. & Northcote, D.M. (1980). The purification of potato lectin by affinity chromatography on a fetuin-Sepharose matrix. *Phytochemistry*, 19, 1861–2.

Paine, P.L., Moore, L.C. & Horovitz, S.B. (1975). Nuclear envelope permeability. *Nature, London*, 254, 109–14.

Painter, R.G. & Ginsberg, M. (1982). Concanavalin A induces interactions between surface glycoproteins and the platelet cytoskeleton. *The Journal of Cell Biology*, 92, 565–73.

Palmer, R.M., Pusztai, A., Bain, P. & Grant, G. (1987). Changes in rates of tissue protein synthesis in rats induced *in vivo* by consumption of kidney bean lectins. *Comparative Biochemistry and Physiology*, 88C, 179–83.

Pastan, I.H. & Willingham, M.C. (1981). Journey to the centre of the cell: Role of the receptosome. *Science*, 214, 504–9.

Pastan, I., Willingham, M.C. & FitzGerald, D.J.P. (1986). Immunotoxins. *Cell*, 47, 641–8.

Patel, J. & Kassis, S. (1987). Concanavalin A prevents phorbol-mediated redistribution of protein kinase C and beta-adrenergic receptors in rat glioma C6 cells. *Biochemical and Biophysical Research Communications*, 144, 1265–72.

Patton, W.F., Dhanak, M.R. & Jacobson, B.S. (1969). Differential partitioning of plasma membrane proteins into the Triton X-100-insoluble cytoskeleton fraction during concanavalin A-induced receptor redistribution. *Journal of Cell Science*, 92, 85–91.

Paul, I. & Devor, M. (1987). Completeness and selectivity of ricin 'suicide transport' lesion in rat dorsal root ganglia. *Journal of Neuroscience Methods*, 22, 103–11.

Pereira, M.E.A., Kabat, E.A. & Sharon, N. (1974). Immunochemical studies on the specificity of soybean agglutinin. *Carbohydrate Research*, 37, 89–102.

Pereira, M.E.A., Kisailus, E.C., Gruezo, F.G. & Kabat, E.A. (1978). Immunochemical studies on the combining site of the blood group H-specific Lectin I from *Ules europeus* seeds. *Archives of Biochemistry and Biophysics*, 185, 108–15.

Pereira, M.E.A., Gruezo, F. & Kabat, E.A. (1979). Purification and characterization of Lectin II from *Ulex europeus* seeds and an immunochemical study of its combining site. *Archives of Biochemistry and Biophysics*, 194, 511–25.

Perkkio, M., Savilahti, E. & Knitunen, P. (1981). Morphometric and immunohistochemical study of jejunal biopsies from children with intestinal soy allergy. *European Journal of Pediatrics*, 137, 63–9.

Peumans, W.J., De Ley, M. & Broekaert, W.F. (1984). An unusual lectin from stinging nettle (*Urtica dioica*) rhizomes. *FEBS Letters*, 177, 99–103.

Peumans, W.J., Nsimba-Lubaki, M., Peeters, B. & Broekaert, W.F. (1985). Isolation and partial characterization of a lectin from ground elder (*Aegopodium podagraria*) rhizomes. *Planta*, 164, 75–82.

Pierce, F.L. (1981). Characterization of and calcium requirement for histamine release from rat peritoneal mast cells treated with concanavalin A. *International Archives of Allergy and Applied Immunology*, **66**, 68–75.

Pirker, R., FitzGerald, D.J.P., Willingham, M.C. & Pastan, I. (1988). Enhancement of the activity of immunotoxins made with either ricin A chain or *Pseudomonas* exotoxin in human ovarian and epidermoid carcinoma cell lines. *Cancer Research*, **48**, 3919–23.

Pistole, T.G. (1981). Interaction of bacteria and fungi with lectins and lectin-like substances. *Annual Review of Microbiology*, **35**, 85–112.

Poley, J.R. & Klein, A.W. (1983). Scanning electron microscopy of soy protein-induced damage of small bowel mucosa in infants. *Journal of Pediatric Gastroenterology and Nutrition*, **2**, 271–87.

Pongor, S. & Reidl, Z. (1983). A latex agglutination test for lectin binding. *Analytical Biochemistry*, **129**, 51–6.

Poretz, R.D. & Goldstein, I.J. (1971). Protein-carbohydrate interaction. Mode of binding of aromatic moieties to concanavalin A, the phytohemagglutinin in jack bean. *Biochemical Pharmacology*, **20**, 2727–39.

Poretz, R.D., Riss, H., Timberlake, J.W. & Chien, S. (1974). Purification and properties of the hemagglutinin from *Sophora japonica* seeds. *Biochemistry*, **13**, 250–6.

Prince, R.I., Miller, B.G., Bailey, M., Telemo, E., Patel, D. & Bourne, F.J. (1988). An ELISA-technique for the determination of soyabean lectin in animal feeds. *Proceedings of British Society for Animal Production*, paper no. 88.

Pueppke, S.G., Bauer, W.D., Keegstra, K. & Ferguson, A.L. (1978). Role of lectins in plant-microorganism interactions. *Plant Physiology*, **61**, 779–84.

Purrello, F., Burnham, D.B. & Goldfine, I.D. (1983a). Insulin receptor antiserum and plant lectins mimic the direct effects of insulin on nuclear envelope phosphorylation. *Science*, **221**, 462–4.

—(1983b). Insulin regulation of protein phosphorylation in isolated rat liver nuclear envelopes: Potential relationship to mRNA metabolism. *Proceedings of the National Academy of Sciences, USA*, **80**, 1189–93.

Pusztai, A. (1964). Hexosamines in the seeds of higher plants (Spermatophytes). *Nature*, London, **201**, 1328–9.

—(1980). Nutritional toxicity of the kidney bean (*Phaseolus vulgaris*). *Annual Report of the Rowett Research Institute*, **36**, 110–18.

—(1985). Constraints on the nutritional utilization of plant proteins. *Nutrition Abstracts & Reviews* (ser. B), **55**, 363–9.

—(1986a). The role in food poisoning of toxins and allergens from higher plants. In *Developments in Food Microbiology* (Robinson, R.K. ed.), pp. 179–94. Elsevier Applied Science Publishers, London & New York.

—(1986b). The biological effects of lectins in the diet of animals and man. In *Lectins, Biology, Biochemistry and Clinical Biochemistry* (Bog-Hansen, T.C. & van Driessche, E. eds), pp. 317–27. Walter de Gruyter, Berlin & New York.

—(1989a). Lectins. In *Toxicants of Plant Origin* vol. III (Cheeke, P.R. ed.), pp. 29–71. CRC Press Inc., Boca Raton, Flo.

—(1989b). Effects on gut structure, function and metabolism of dietary lectins. The nutritional toxicity of the kidney bean lectin. In *Advances in Lectin Research* (Franz, H. ed.), pp. 74–86. VEB Verlag Volk und Gesundheit, Berlin.

—(1989c). Transport of proteins through the membranes of the adult gastrointestinal tract – a potential for drug delivery? *Advanced Drug Delivery Reviews*, **3**, 215–28.

Pusztai, A. & Greer, F. (1984). Effects of dietary legume proteins on the morphology and secretory responses of the rat small intestine. *Protides of the Biological Fluids*, **32**, 347–50.

Pusztai, A. & Palmer, R. (1977). Nutritional evaluation of kidney bean (*Phaseolus vulgaris*): the toxic principle. *The Journal of Science of Food and Agriculture*, **28**, 620–3.

Pusztai, A. & Stewart, J.C. (1978). Isolectins of *Phaseolus vulgaris*. Physicochemical studies. *Biochimica et Biophysica Acta*, **536**, 38–49.

Pusztai, A. & Watt, W.B. (1974). Isolectins of *Phaseolus vulgaris*. A comprehensive study of fractionation. *Biochimica et Biophysica Acta*, **365**, 57–71.

Pusztai, A., Clarke, E.M.W. & King, T.P. (1979*b*). The nutritional toxicity of *Phaseolus vulgaris* lectins. *Proceedings of the Nutritional Society*, **38**, 115–20.

Pusztai, A., Grant, G. & Oliveira, de, J.T.A., (1986). Local (gut) and systemic responses to dietary lectins. *IRCS Medical Science*, **14**, 205–8.

Pusztai, A., Grant, G. & Palmer, R. (1975). Nutritional evaluation of kidney beans (*Phaseolus vulgaris*): the isolation and partial characterization of toxic constituents. *The Journal of Science of Food and Agriculture*, **20**, 149–56.

Pusztai, A., Grant, G. & Stewart, J.C. (1981*a*). A new type of *Phaseolus vulgaris* (cv. Pinto 111) seed lectin: isolation and characterization. *Biochimica et Biophysica Acta*, **671**, 146–54.

—(1982*a*). The isolation and characterization of an unusual seed lectin from a 'lectin-free' cultivar of *Phaseolus vulgaris*, Pinto 111, and its relationship to lectins synthesized by root cells. In *Lectins, Biology, Biochemistry and Clinical Biochemistry* (Bog-Hansen, T.C. ed.), vol. 2, pp. 743–758. Walter de Gruyter, Berlin and New York.

Pusztai, A., Greer, F. & Grant, G. (1989*b*). Specific uptake of dietary lectins into the systemic circulation of rats. *Biochemical Society Transactions*, **17**, 481–2.

Pusztai, A., King, P.T. & Clarke, E.M.W. (1982*b*). Recent advances in the study of nutritional toxicity of the kidney bean (*Phaseolus vulgaris*) lectins in rats. *Toxicon*, **20**, 195–7.

Pusztai, A., Stewart, J.C. & Watt, W.B. (1978). A novel method for the preparation of protein bodies by filtration in high (over 70% w/v) sucrose-containing media. *Plant Science Letters*, **12**, 9–15.

Pusztai, A., Watt, W.B. & Stewart, J.C. (1987). Erythro- and lympho-agglutinins of *Phaseolus acutifolius* L. *Phytochemistry*, **26**, 1009–13.

—(1991). A comprehensive scheme for the isolation of trypsin inhibitors and the agglutinin from soybean seeds. *Journal of Agricultural and Food Chemistry*, **39**, 862–6.

Pusztai, A., Clarke, E.M.W., Grant, G. & King, T.P. (1981*b*). The toxicity of *Phaseolus vulgaris* lectins. Nitrogen balance and immunochemical studies. *The Journal of the Science of Food and Agriculture*, **32**, 1037–46.

Pusztai, A., Croy, R.R.D., Grant, G. & Watt, W.B. (1977). Compartmentalization in the cotyledonary cells of *Phaseolus vulgaris* L. seeds: a differential sedimentation study. *The New Phytologist*, **79**, 61–71.

Pusztai, A., Croy, R.R.D., Stewart, J.C. & Watt, W.B. (1979*a*). Protein body membranes of *Phaseolus vulgaris* L. cotyledons: isolation and preliminary characterization of constituent proteins. *The New Phytologist*, **83**, 371–8.

Pusztai, A., Croy, R.R.D., Grant, G. & Stewart, J.C. (1983*a*). Seed lectins: Distribution, location and biological role. In *Seed Proteins* (Daussant, J., Mosse, J. & Vaughan, J. eds), pp. 53–82. Academic Press, New York.

Pusztai, A., Greer, F., Silva Lima, M. de G., Prouvost-Danon, A. & King, T.P. (1983*b*). Local and systemic responses to dietary lectins. In *Chemical Taxonomy, Molecular Biology and Function of Plant Lectins* (Etzler, M.E. & Goldstein, I.J. eds), pp. 271–2. Alan R. Liss, Inc. New York.

Pusztai, A. Oliveira, de, J.T.A., Bardocz, S., Grant, G. & Wallace, H.M. (1988*a*). Dietary kidney bean lectin-induced hyperplasia and increased polyamine content of the small intestine. In *Lectins, Biology, Biochemistry and Clinical Biochemistry*, (Bog-Hansen, T.C. & Freed, D.L.J. eds), vol. 6, pp. 117–20. Sigma Library, St. Louis, Mo.

Pusztai, A., Grant, G., Brown, D.S., Ewen, S.W.B. & Bardocz, S. (1988*b*). *Phaseolus vulgaris* lectin induces growth and increases the polyamine content of rat small intestine *in vivo*. *Medical Science Research*, **16**, 1283–4.

Pusztai, A., Grant, G., Williams, L.M., Brown, D.S., Ewen, S.W.B. & Bardocz, S. (1989*a*). *Phaseolus vulgaris* lectin induces growth and the uptake of polyamines by the rat small intestine *in vivo*. *Medical Science Research*, **17**, 143–5.

Pusztai, A., Ewen, S.W.B., Grant, G., Peumans, W.J., van Damme, E.J.M., Rubio, L. & Bardocz, S. (1990*a*). The relationship between survival and binding of plant lectins during small intestinal passage and their effectiveness as growth factors. *Digestion*, **46** (suppl. 2) 308–16.

Pusztai, A., Grant, G., King, T.P. & Clarke, E.W.M. (1990*b*). Chemical probiosis. In *Recent Advances in Animal Nutrition*, Proceedings of Food Manufacturers Conference, Nottingham, 1990 (Haresign, W. & Cole, D.J.A. eds). Butterworth Scientific, London, in press.

Quinn, J.M. & Etzler, M.E. (1987). Isolation and characterization of a lectin from the roots of *Dolichos biflorus*. *Archives of Biochemistry and Biophysics*, **258**, 535–44.

Raikhel, N.V. & Quatrano, R.S. (1986). Localization of wheat germ agglutinin in developing wheat embryos and those cultured in abscisic acid. *Planta*, **168**, 433–40.

Raikhel, N.V. & Wilkins, T.A. (1987). Isolation and characterization of a cDNA clone encoding wheat germ agglutinin. *Proceedings of the National Academy of Sciences, USA*, **84**, 6745–9.

Raikhel, N.V., Palevitz, B.A. & Haigler, C.H. (1986). Abscisic acid control of lectin accumulation in wheat seedlings and callus cultures. *Plant Physiology*, **80**, 167–71.

Rainer, O. (1962). Zur Vergiftung mit rohen, grünen Bohnen (Phasin Vergiftung). *Medizinische Klinik*, **57**, 270–2.

Rapin, A.M.C. & Burger, M. (1974). Tumor cell surfaces. General alterations detected by agglutinins. *Advances in Cancer Research*, **20**, 1–91.

Rattray, E.A.S., Palmer, R. & Pusztai, A. (1974). Toxicity of kidney beans (*Phaseolus vulgaris* L.) to conventional and gnotobiotic rats. *The Journal of the Science of Food and Agriculture*, **25**, 1035–40.

Read, J.C., Robb, R.J., Greene, W.C. & Nowell, P.C. (1985). Effect of wheat germ agglutinin on the interleukin pathway of the human T lymphocyte activation. *The Journal of Immunology*, **134**, 314–23.

Reinherz, E.L., Kung, P.C., Goldstein, G. & Shlossman, S.F. (1979). Separation of functional subsets of human T cells by a monoclonal antibody. *Proceedings of the National Academy of Sciences, USA*, **76**, 4061–5.

Reisner, Y., Sharon, N. & Haran-Ghera, N. (1980). Expression of peanut agglutinin receptors on virus-induced pre-leukemic cells in mice. *Proceedings of the National Academy of Sciences, USA*, **77**, 2244–6.

Renkonnen, K.O. (1948). Studies on hemagglutinins present in seeds of some representatives of Leguminoseae. *Annales Medicinae Experimentalis et Biologiae Fenniae*, **26**, 66–72.

—(1960). The development of hemagglutinins in the seeds of *Vicia cracca*. *Annales Medicinae Experimentalis et Biologiae Fenniae*, **38**, 26–9.

Restum-Miguel, N. & Prouvost-Danon, A. (1985). Effects of multiple oral dosing on IgE synthesis in mice: oral sensitization by albumin extracts from seeds of Jack fruit (*Artocarpus integrifolia*) containing lectins. *Immunology*, **54**, 497–504.

Rice, L., Laughter, A.H. & Twomey, J.J. (1979). Three suppressor systems in human blood that modulate lymphoproliferation. *The Journal of Immunology*, **122**, 991–6.

Rich, R.R. & Pierce, C.W. (1974). Biologic expressions of lymphocyte activation. III. Suppression of plaque-forming cell responses in vitro by supernatant fluids from concanavalin A-activated spleen cell cultures. *The Journal of Immunology*, **112**, 1360–8.

Richardson, M., Campos, F.D.A.P., Moreira, R.A., Ainouz, I.L., Begbie, R., Watt, W.B. & Pusztai, A. (1984). The complete amino acid sequence of the major α-subunit of the lectin from the seeds of *Dioclea grandiflora* (Mart.). *European Journal of Biochemistry*, **144**, 101–11.

Rigas, D.A. & Osgood, E.A. (1955). Purification and properties of the phytohemagglutinin of *Phaseolus vulgaris*. *The Journal of Biological Chemistry*, **212**, 607–15.

Rittman, B.R., Mackenzie, I.C. & Rittman, G.A. (1982). Lectin binding to murine oral mucosa and skin. *Archives of Oral Biology*, **27**, 1013–19.

Roberts, D.D. & Goldstein, I.J. (1983*a*). Binding of hydrophobic ligands to plant lectins: Titration with arylaminonaphthalenesulfonates. *Archives of Biochemistry and Biophysics*, **224**, 479–84.

—(1983*b*). Adenine binding sites of the lectin from lima beans (*Phaseolus lunatus*). *Journal of Biological Chemistry*, **258**, 13820–40.

Roberts, L.M. & Lord, J.M. (1981). The synthesis of *Ricinus communis* agglutinin. Cotranslational and posttranslational modification of agglutinin polypeptides. *European Journal of Biochemistry*, **119**, 31–41.

Rolfe, B.G., Redmond, J.W., Batley, M., Chen, H., Djordjevic, S.P., Ridge, R.W., Bassam, B.J., Sargent, C.L., Dazzo, R.B. & Djordevic, M.A. (1986). Intercellular communication and recognition in the Rhizobium-legume symbiosis. In *Recognition in Microbe–Plant Symbiotic and Pathogenic Interactions* (Lugtenberg, B. ed.), pp. 39–54. Springer-Verlag, Berlin.

Rossi, M.A., Mancini Filho, J., Jr & Lajolo, F.M. (1984). Jejunal ultrastructural changes induced by kidney bean (*Phaseolus vulgaris*) lectins in rats. *The British Journal of Experimental Pathology*, **65**, 117–23.

Roth, R.A., Cassell, D.J., Maddux, B.A. & Goldfine, T.D. (1983). Regulation of insulin receptor kinase activity by insulin mimickers and an insulin antagonist. *Biochemical and Biophysical Research Communications*, **115**, 245–52.

Rouanet, J.M. & Besancon, P. (1979). Effets d'un extrait de phytohemagglutinines sur la croissance, la digestibilité de l'azote et l'activité de l'invertase et de la (Na+-K+)-ATP-ase de la muquese intestinale chez le rat. *Annales de la Nutrition et de l'Alimentation*, **33**, 405–16.

Rouanet, J.M., Besancon, P. & Lafont, J. (1983). Effect of lectins from leguminous seeds on rat duodenal enterokinase activity. *Experientia*, **39**, 1356–8.

Rouanet, J.M., Lafont, J., Zambonino-Infante, J.L. & Besancon, P. (1988). Selective effects of PHA on rat brush border hydrolases along the crypt-villus axis. *Experientia*, **44**, 340–1.

Rouge, P., Richardson, M., Ranfaing, P., Yarwood, A. & Sousa-Cavada, B. (1987). Single- and two-chain legume lectins as phylogenetic markers of speciation. *Biochemical Systematics and Ecology*, **15**, 341–8.

Ruda, M. & Coulter, J.D. (1982). Axonal and transneuronal transport of wheat germ agglutinin demonstrated by immunocytochemistry. *Brain Research*, **249**, 237–46.

Rumsby, G. & Puck, T.T. (1982). Ornithine decarboxylase and the cytoskeleton in normal and transformed cells. *Journal of Cellular Physiology*, **111**, 133–9.

Rutherford, W.M., Dick, W.E. Jr, Cavins, J.F., Dombrink-Kurtzman, M.A. & Slodki, M.E. (1986). Isolation and characterization of a soybean lectin having 4-o-methylglucuronic acid specificity. *Biochemistry*, **25**, 952–8.

Sabnis, D.D. & Hart, J.W. (1978). The isolation and some properties of a lectin (haemagglutinin) from *Cucurbita* phloem exudate. *Planta*, **142**, 97–101.

Saito, M., Takuku, F., Hayashi, M., Tanaka, I., Abe, Y., Nagai, Y. & Ishii, S. (1983). The role of valency of concanavalin A and its chemically modified derivatives in lymphocyte activation. Monovalent monomeric concanavalin A derivative can stimulate lymphocyte blastoid transformation. *The Journal of Biological Chemistry*, **258**, 7499–505.

Salvaterra, P.M., Gurd, J.M., Mahler, H.R. (1977). Interaction of the nicotinic acetylcholine receptor from rat brain with lectins. *Journal of Neurochemistry*, **29**, 345–8.

Salyajev, R.K. & Kuzevanov, V.Y. (1984). Lectin receptors on the tonoplast and agglutination of isolated vacuoles. *Fiziologia Rastenii*, **31**, 73–81.

Sampson, D., Grotelueschen, C. & Kauffman, H.M. Jr (1975). The human splenic suppressor cell. *Transplantation*, **20**, 362–7.

Santoro, L.G., Grant, G. & Pusztai, A. (1988). Differences in the degradation *in vivo* and *in vitro* of phaseolin, the major storage protein of *Phaseolus vulgaris* seeds. *Biochemical Society Transactions*, **16**, 612–13.

Schechter, Y. (1983). Bound lectins that mimic insulin produce persistent insulin-like activities. *Endocrinology*, **113**, 1921–6.

Schechter, B., Lis, H., Lotan, R., Novogrodsky, A. & Sharon, N. (1976). The requirement for tetravalency of soybean agglutinin for induction of mitogenic stimulation of lymphocytes. *European Journal of Immunology*, **6**, 145–9.

Schlessinger, J., Elson, E.L., Webb, W.W., Yakara, I., Rutischauser, U. & Edelman, G.M. (1977). Receptor diffusion on cell surfaces modulated by locally bound concanavalin A. *Proceedings of the National Academy of Sciences, USA*, **74**, 1110–14.

Schmidt, M.L. & Trojanowski, J.Q. (1985). Immunoblot analysis of horse-radish peroxidase conjugates of wheat germ agglutinin before and after retrograde transport in the rat peripheral nervous system. *The Journal of Neuroscience*, **5**, 2779–85.

Schnell, D.J. & Etzler, M.E. (1987). Primary structure of the *Dolichos biflorus* seed lectin. *The Journal of Biological Chemistry*, **262**, 7720–5.

—(1988). cDNA cloning, primary structure and *in vitro* biosynthesis of the DB58 lectin from *Dolichos biflorus*. *The Journal of Biological Chemistry*, **263**, 14648–53.

Schnell, D.J., Alexander, D.C., Williams, B.G. & Etzler, M.E. (1987). cDNA cloning and *in vitro* synthesis of the *Dolichos biflorus* seed lectin. *European Journal of Biochemistry*, **167**, 227–31.

Schoonhoven, A.V., Cardona, C. & Valor, J. (1983). Resistance to bean weevil and the Mexican bean weevil (Coleoptera: Bruchidae) in noncultivated bean accessions. *Journal of Economy and Entomology*, **76**, 1255–9.

Schreiber, A.B., Liberman, T.A., Lax, I., Yarden, Y. & Schlessinger, J. (1983). Biological role of epidermal growth factor-receptor clustering: Investigation with monoclonal anti-receptor antibodies. *The Journal of Biological Chemistry*, **258**, 846–53.

Seidel, E.R., Haddox, M.K. & Johnson, L.R. (1984). Polyamines in the response to intestinal obstruction. *The American Journal of Physiology*, **246**, G649–53.

Seidel, E.R., Tabata, K., Dembinski, A.B. & Johnson, L.R. (1985). Attenuation of trophic response to gastrin after inhibition of ornithine decarboxylase. *The American Journal of Physiology*, **249**, G16–20.

Seiler, N. (1987). Functions of polyamine acetylations. *Canadian Journal of Physiology and Pharmacology*, **65**, 2024–35.

Sekine, I., Kawase, Y., Nishimori, I., Mitarai, M., Harada, H., Ishiguro, M. & Kikutani, M. (1986). Pathological study on mucosal changes in small intestine of rat by oral administration of ricin. I. Microscopical observations. *Acta Pathologica Japan*, **36**, 1205–12.

Sequeira, L. (1978). Lectins and their role in host-pathogen specificity. *Annual Review of Phytopathology*, **16**, 453–81.

—(1984). Recognition systems in plant-pathogen interactions. *Biology of the Cell*, **51**, 281–6.

Sequeira, L. & Graham, T.L. (1977). Agglutination of avirulent strains of *Pseudomonas solanacearum* by potato lectin. *Physiology and Plant Pathology*, **11**, 43–54.

Sequeira, L., Gaard, G. & De Zoeten, G.A. (1977). Attachment of bacteria to host cell walls: Its relation to mechanisms of induced resistance. *Physiology and Plant Pathology*, **10**, 43–50.

Sharon, N. (1984). Surface carbohydrates and surface lectins are recognition determinants in phagocytosis. *Immunology Today*, **5**, 143–7.

Sharon, N. (1987). Bacterial lectins, cell–cell recognition and infectious disease. *FEBS Letters*, **217**, 145–57.

Shibuya, N., Goldstein, I.J., Broekart, W.F., Nsimba-Lubaki, M., Peeters, B. & Peumans, W.J. (1987). The elderberry (*Sambucus nigra* L.) bark lectin recognizes the Neu5Ac(2-6)Gal/GalNAc sequence. *The Journal of Biological Chemistry*, **262**, 1596–601.

Shimoda, T. & Funatsu, G. (1985). Binding of lactose and galactose to native and iodinated ricin D. *Agricultural and Biological Chemistry*, **49**, 2125–30.

Shirley, S.G., Polak, E.H., Mather, R.A. & Dodd, G.H. (1987*a*). The effect of concanavalin A on the rat electro-olfactogram. Differential inhibition of odorant response. *The Biochemical Journal*, **245**, 175–84.

Shirley, S.G., Polak, E.H., Edwards, D.A., Wood, M.A. & Dodd, G.H. (1987*b*). The effect of concanavalin A on rat electro-olfactogram at various odorant concentrations. *The Biochemical Journal*, **245**, 185–9.

Shukle, R.H. & Murdoch, L.L. (1983). Lipoxygenase, trypsin inhibitor and lectin from soybeans, effects on larval growth of *Manduca sexta* (Lepidoptera-Sphingidae). *Environmental Entomology*, **12**, 787–91.

Simerly, R.B. & Swanson, L.W. (1988). Projections of the medial preoptic nucleus: A *Phaseolus vulgaris* agglutinin anterograde tract-tracing study in the rat. *The Journal of Comparative Neurology*, **270**, 209–42.

Simmons, B.M., Stahl, P.D. & Russell, J.H. (1986). Mannose-receptor mediated uptake of ricin toxin and ricin A chain by macrophages. Multiple intracellular pathways for A chain translocation. *The Journal of Biological Chemistry*, **261**, 7912–20.

Sing, V.O. & Schroth, M.N. (1977). Bacteria–plant cell surface interactions: Active immobilization of saprophytic bacteria in plant leaves. *Science*, **197**, 759–61.

Siraganian, P.A. & Siraganian, R.P. (1974). Basophil activation by concanavalin A. Characteristics of the reaction. *The Journal of Immunology*, **112**, 2117–25.

Sitkovsky, M.V., Pasternack, M.S. & Eisen, H.N. (1982). Inhibition of cytotoxic T lymphocyte activity by concanavalin A. *The Journal of Immunology*, **129**, 1372–6.

Sjölander, A. (1988). Direct effects of wheat germ agglutinin on inositol phosphate formation and cytosolic free calcium level in intestine 407 cells. *Journal of Cellular Physiology*, **134**, 473–8.

Sjölander, A., Magnusson, K.E. & Latkovic, S. (1984). The effect of concanavalin A and wheat germ agglutinin on the ultrastructure and permeability of the small intestine. A possible model for an intestinal allergic reaction. *International Archives of Allergy and Applied Immunology*, **75**, 230–6.

—(1986). Morphological changes of rat small intestine after short-time exposure to concanavalin A or wheat germ agglutinin. *Cell Structure and Function*, **11**, 285–93.

Skilleter, D.N., Price, R.J. & Thorpe, P.E. (1985). Modification of the carbohydrate in ricin with metaperiodate and cyanoborohydride mixtures: effect on binding, uptake and toxicity to parenchymal and non-parenchymal cells of rat liver. *Biochimica et Biophysica Acta*, **842**, 12–21.

Skoog, V.T., Nilsson, S.F. & Weber, T.H. (1980). Characterization of human lymphocyte surface receptors for mitogenic and non-mitogenic substances. *Scandinavian Journal of Immunology*, **11**, 369–76.

Skubatz, H. & Kessler, B. (1984). A development-dependent hemagglutinin from cucumber surfaces. *Plant Physiology*, **76**, 55–8.

Slama, J.S. & Rando, R.R. (1980). Lectin-mediated aggregation of liposomes containing glycolipids with variable hydrophylic spacer arms. *Biochemistry*, **19**, 4595–600.

Smit, G., Kijne, J.W. & Lugtenberg, B.J.J. (1986). Correlation between extracellular fibrils and attachment of *Rhizobium leguminosarum* to pea root hair tips. *The Journal of Bacteriology*, **168**, 821–7.

—(1987). Involvement of both cellulose fibrils and Ca^{2+}-dependent adhesin in the attachment of *Rhizobium leguminosarum* to pea root hair tips. *The Journal of Bacteriology*, **169**, 4294–301.

Smith, J.D. & Liu, A.Y.C. (1981). Lectins mimic insulin in induction of tyrosine aminotransferase. *Science*, **214**, 799–800.

Smith, L.M., Sabnis, D.D. & Johnson, R.P.C. (1987). Immunocytochemical localization of phloem lectin from *Cucurbita maxima* using peroxidase and colloidal-gold labels. *Planta*, **170**, 461–70.

Smith, S.R., Umland, S., Terminelli, C. & Watnick, A.S. (1984). A study of the mechanism of Con A-induced immunosuppression *in vivo*. *Cellular Immunology*, **87**, 147–58.

Sollid, L.M., Kolberg, J., Scott, H., Ek, J., Fausa, O. & Brandtzaeg, P. (1986). Antibodies to wheat germ agglutinin in coeliac disease. *Clinical and Experimental Immunology*, **63**, 95–100.

Sorimachi, K. (1984). Temperature- and dose-dependent internalization of concanavalin A in monolayer culture. *Biochemical and Biophysical Research Communications*, **125**, 35–44.

Sorimachi, K., Okayasu, T. & Yasumura, Y. (1987). Concanavalin A changes not only the number of insulin-binding sites but also the binding affinity in rat hepatoma cells in culture. *Endocrine Research*, **13**, 183–97.

Sotelo, A., Arteaga, M.E., Frias, M.I. & Gonzales-Garza, M.T. (1980). Cytotoxic effect of two legumes in epithelial cells of the small intestine. *Qualitas Plantarum. Plant Foods for Human Nutrition*, **30**, 79–85.

Sotelo, A., Licea, A.G., Gonzales-Garza, M.T., Velasco, E. & Feria-Velasco, A. (1983). Ultrastructural changes of epithelial intestinal cells induced by the ingestion of raw *Phaseolus acutifulius*. *Nutrition Reports International*, **27**, 329–37.

Southworth, D. (1975). Lectins stimulate pollen germination. *Nature*, London, **258**, 600–2.

Spadoro-Tank, J.P. & Etzler, M.E. (1988). Heat shock enhances the synthesis of a lectin-related protein in *Dolichos biflorus* cell suspension cultures. *Plant Physiology*, **88**, 1131–5.

Spitler, L.E., del Rio, M., Kheuntigan, A., Wedel, N.I., Brody, N.A., Miller, L.L., Harkonen, W.S., Rosendorf, L.L., Lee, H.M., Mischak, R., Kawahata, R.T., Stoudemire, J.B., Fradkin, L.B., Bantista, E.E. & Scannon, P.J. (1987). Therapy of patients with malignant melanoma using monoclonal anti-melanoma antibody-ricin A chain immunotoxin. *Cancer Research*, **47**, 1717–23.

Stacey, G., Halverson, L.J., Nieuwkoop, T., Banfalvy, Zs., Schell, M.G., Gerhold, D., Deshamene, N., So, J.S. & Sirotkin, K.M. (1986). Nodulation of soybean: *Bradyrhizobium japonicum* physiology and genetics. In *Recognition in Microbe-Plant Symbiotic and Pathogenic Interactions* (Lugtenberg, B. ed.), pp. 87–99. Springer-Verlag, Berlin.

Stern, M. & Gellerman, B. (1988). Food proteins and maturation of small intestinal microvillus membranes (MVM) I. Binding characteristics of cow's milk proteins and concanavalin A to MVM from newborn and adult rats. *Journal of Pediatric Gastroenterology and Nutrition*, **7**, 115–21.

Stern, M., Gellerman, B., Belitz, H.D. & Wieser, H. (1988). Food proteins and maturation of the small intestinal microvillus membranes (MVM). II. Binding of gliadin hydrolysate fractions and of the gliadin peptide B 3142. *Journal of Pediatric Gastroenterology and Nutrition*, **7**, 122–7.

Stillmark, H. (1888). Über Ricin ein giftiges ferment aus den samen von *Ricinus communis* L. und einigen anderen Euphorbiaceen. *Inaugural Dissertation Dorpat (Tartu)*.

—(1889). Über Ricin. In *Arbeiten des Pharmakologischen Instituts zu Dorpat*. (Kobert, R. ed.), pp. 59–151. Enke, Stuttgart.

Stinissen, H.M., Peumans, W.J. & Carlier, A.R. (1982). *In vivo* synthesis and processing of cereal lectins. *Plant Molecular Biology*, **1**, 277–90.

—(1983). Two-step processing of *in vivo* synthesized rice lectin. *Plant Molecular Biology*, **2**, 33–40.

Stinissen, H.M., Peumans, W.J. & Chrispeels, M.J. (1985). Posttranslational processing of proteins in vacuoles and protein bodies is inhibited by monensin. *Plant Physiology*, **77**, 495–8.

Stirpe, F. & Barbieri, L. (1986). Ribosome-inactivating proteins up to date. *FEBS Letters*, **195**, 1–8.

Stone-Wolff, D.S., Yip, Y.K., Kelker, H.C., Le, J., Henriksen-Destefano, D., Rubin, B.Y., Rinderknecht, E., Aggarwal, B.B. & Vilcek, J. (1984). Interrelationships of human interferon-gamma with lymphotoxin and monocyte cytotoxin. *The Journal of Experimental Medicine*, **159**, 828–43.

Strazza, S. & Sherry, A.D. (1982). Concanavalin A will not assume the sugar binding conformation in the complete absence of metal ions. *Biochemical and Biophysical Research Communications*, **106**, 1291–7.

Strobel, G.A. (1974). The toxin-binding protein of sugarcane, its role in the plant and in disease development. *Proceedings of the National Academy of Sciences, USA*, **71**, 4232–6.

Strosberg, A.D., Lauweyers, M. & Foriers, A. (1983). Molecular evolution of legume lectins. In *Chemical Taxonomy. Molecular Biology and Function of Plant Lectins* (Goldstein, I.J. & Etzler, M.R. eds), pp. 7–20. Alan R. Liss Inc., New York.

Strosberg, A.D., Buffard, D., Lauwereys, M. & Foriers, A. (1986a). Legume lectins: A large family of homologous proteins. In *The Lectins* (Liener, I.E., Sharon, N. & Goldstein, I.J. eds), pp. 249–64. Academic Press, Orlando.

Strosberg, A.D., Buffard, D., Kaminski, A.P., Chapot, M.P., Rossow, P.W. & Foriers, A. (1986b). Lectin multigene families in leguminous and nonleguminous plants. In *Molecular Biology of Seed Storage Proteins and Lectins* (Shannon, L.W. & Chrispeels, M.J. eds) pp. 1–28. American Society of Plant Physiologists, Warverley Press, Baltimore.

Sugiura, Y., Lee, C.L. & Perl, E.R. (1986). Central projections of identified unmyelinated (C) afferent fibers innervating mammalian skin. *Science*, **234**, 358–61.

Sula, K., Paluska, E., Nouza, K. & Danek, P.F. (1986). Lentil lectin inhibits cells producing graft-versus-host reaction but does not suppress hematopoietic stem cells in mice. *Immunobiologia*, **173**, 35–40.

Sumner, J.B. & Howell, S.F. (1936). The identification of the hemagglutinin of the jack bean with concanavalin A. *The Journal of Bacteriology*, **32**, 227–37.

Tabata, K. & Johnson, L.R. (1986). Ornithine decarboxylase and mucosal growth in response to feeding. *The American Journal of Physiology*, **251**, G270–4.

Tajiri, H., Lee, P.C. & Lebenthal, E. (1986). Small intestinal hyperplasia caused by an enterokinase inhibitor from red kidney bean in rats. *The Journal of Nutrition*, **116**, 873–80.

Talbot, C.F. & Etzler, M.E. (1978). Isolation and characterization of a protein from the leaves and stems of *Dolichos biflorus* that cross reacts with antibodies to the seed lectin. *Biochemistry*, **17**, 1474–9.

Tartakoff, A.M. & Vassali, P. (1983). Lectin-binding sites as markers of Golgi subcompartments: proximal to distal maturation of oligosaccharides. *The Journal of Cell Biology*, **97**, 1243–8.

Taylor, M.V., Hesketh, T.R. & Metcalf, J.C. (1988). Phosphoinositide metabolism and the calcium response to concanavalin A in S49 T-lymphoma cells. A comparison with thymocytes. *The Biochemical Journal*, **249**, 847–55.

Thorpe, P.E. (1985). Antibody carriers of cytotoxic agents in cancer therapy: a review. In *Monoclonal Antibodies '84: Biological and Clinical Applications* (Pinchera, A., Doria, G., Dammacco, F. & Bargellesi, A. eds), pp. 475–506.

Thorpe, P.E., Cumber, A.J., Williams, N., Edwards, D.C., Ross, W.C.J. & Davis, A.J.S. (1981). Abrogation of the non-specific toxicity of abrin conjugated to anti-lymphocyte globulin. *Clinical and Experimental Immunology*, **43**, 195–200.

Thorpe, P.E., Ross, W.C.J., Brown, A.N.F., Myers, C.D., Cumber, A.J., Foxwell, B.M.J. & Forrester, J.T. (1984). Blockade of galactose-binding sites of ricin by its linkage to antibody. *European Journal of Biochemistry*, **140**, 63–71.

Thorpe, P.E., Wallace, P.M., Knowles, P.P., Relf, M.G., Brown, A.N., Watson, G.J., Blakey, D.C. & Newell, D.R. (1983). Improved antitumor effects of immunotoxins prepared with deglycosylated ricin A-chain and hindered disulfide linkages. *Cancer Research*, **48**, 6396–403.

Thorpe, P.E., Wallace, P.M., Knowles, P.P., Relf, M.G., Brown, A.N., Watson, G.J., Knyba, R.E., Wawrzynczak, E.J. & Blakey, D.C. (1987). New coupling agents for the synthesis of immunotoxin containing a hindered disulfide bond with improved stability *in vivo*. *Cancer Research*, **47**, 5924–31.

Tomita, M., Kurokawa, T., Onozaki, K., Ichiki, N., Osawa, T. & Ukita, T. (1972). Purification of galactose-binding phytoagglutinins and phytotoxin by affinity column chromatography using Sepharose. *Experientia*, **28**, 84–5.

Torres-Pinedo, R. (1983). Lectins and the intestine. *The Journal of Pediatric Gastroenterology and Nutrition*, **2**, 588–94.

Toyoshima, S., Hirata, F., Iwata, M., Axelrod, J., Osawa, T. & Waxdal, M.J. (1982). Lectin-induced mitosis and phospholipid methylation. *Molecular Immunology*, **19**, 467–75.

Triadou, N. & Audran, E. (1983). Interaction of the brush border hydrolases of the human small intestine with lectins. *Digestion*, **27**, 1–7.

Triplett, B.A. & Quatrano, R.S. (1982). Timing, localization and control of wheat germ agglutinin synthesis in developing wheat embryos. *Developmental Biology*, **91**, 491–6.

Trojanowski, J.Q. (1983). Native and derivatized lectins for in vivo studies of neuronal connectivity and neuronal cell biology. *Brain Research*, **272**, 201–10.

Truneh, A. & Pierce, F.L. (1981). Characteristics of and calcium requirements for histamine release from rat peritoneal mast cells treated with concanavalin A. *International Journal of Allergy and Applied Immunology*, **66**, 68–75.

Tunis, M., Lis, H. & Sharon, N. (1979). Participation of the carbohydrate moieties of glycolipids and glycoproteins in agglutination of erythrocytes by lectins. *Protides of the Biological Fluids*, **27**, 521–4.

Turner, R.H. & Liener, I.E. (1975). The effect of the selective removal of hemagglutinins on the nutritional value of soybeans. *The Journal of Agriculture and Food Chemistry*, **23**, 484–7.

Ueki, Y., Eguchi, K., Otsubo, T., Kawabe, Y., Shimomura, C., Matsunaga, M., Tezuka, H., Nakao, H., Kawakami, A., Izumi, M., Ishikawa, N., Ito, K. & Nagataki, S. (1988). Phenotypic analyses and concanavalin A-induced suppressor cell dysfunction of intrathyroidal lymphocytes from patients with Graves' disease. *The Journal of Clinical Endocrinology and Metabolism*, **67**, 1018–24.

Ukena, T.C., Goldman, E., Benjamin, T.L. & Karnovsky, M.J. (1976). Lack of correlation between agglutinability, the surface distribution of con A and post-confluence inhibition of cell division in ten cell lines. *Cell*, **7**, 213–22.

Vale, R.D. & Shooter, E.M. (1983). Epidermal growth factor receptors on PC12 cells: Alteration of binding properties by lectins. *Journal of Cell Biochemistry*, **22**, 99–109.

van Damme, E.J.M. & Peumans, W.J. (1987). Isolectin composition of individual clones of *Urtica dioica:* evidence for phenotypic differences. *Physiologia Plantarum*, **71**, 328–34.

van Damme, E.J.M., Allen, A.K. & Peumans, W.J. (1987*a*). Isolation and characterisation of a lectin with exclusive specificity towards mannose from snowdrop (*Galanthus nivalis* L.) bulbs. *FEBS Letters*, **215**, 140–4.

—(1987*b*). Leaves of the orchid twayblade (*Listera ovata*) contain a mannose-specific lectin. *Plant Physiology*, **85**, 566–9.

—(1988*a*). Related mannose specific lectins from different species of Amaryllidaceae. *Physiologia Plantarum*, **73**, 52–7.

van Damme, E.J.M., Broekaert, W.F. & Peumans, W.J. (1988*b*). The *Urtica dioica* agglutinin is a complex mixture of isolectins. *Plant Physiology*, **86**, 598–601.

van Deurs, B., Tonnessen, T.I., Petersen, O.W., Sandvig, K. & Olsnes, S. (1986). Routing of internalized ricin conjugates to the Golgi complex. *Journal of Cell Biology*, **102**, 37–47.

van Deurs, B., Petersen, O.W., Olsnes, S. & Sandvig, K. (1987). Delivery of internalized ricin from endosomes to cisternal Golgi elements is a discontinuous, temperature-sensitive process. *Experimental Cell Research*, **171**, 137–52.

van Deurs, B., Sandvig, K., Petersen, O.W., Olsnes, S., Simons, K. & Griffiths, G. (1988). Estimation of internalized ricin that reaches the *trans*-Golgi network, *The Journal of Cell Biology*, **106**, 253–67.

van Driessche, E. (1987). Structure and function of *Leguminosae* lectins. In *Advances in Lectin Research* (Franz, H. ed.), pp. 73–134. VEB Verlag Volk und Gesundheit, Berlin.

van Dreissche, E., Smets, G., Dejaegere, R. & Kanarek, L. (1981). The immuno-histochemical localization of lectin in pea seeds (*Pisum sativum* L.). *Planta*, **153**, 287–96.

van Ness, B.G., Howard, J.B. & Bodley, J.W. (1980). ADP-rybosylation of elongation factor 2 by diphtheria toxin. Isolation and properties of a novel ribosyl-amino acid and its hydrolysis products. *The Journal of Biological Chemistry*, **255**, 10710–16.

Vilcek, J., Le, J. & Yip, Y.K. (1986). Induction of human interferon gamma with phorbol ester and phytohemagglutinin. *Methods in Enzymology*, **119**, 48–54.

Vitale, A. & Chrispeels, M.J. (1984). Transient N-acetylglucosamine in the biosynthesis of phytohemagglutinin: attachment in the Golgi apparatus and removal in the protein bodies. *The Journal of Cell Biology*, **99**, 133–40.

Vitetta, E.S. & Uhr, J.W. (1984). The potential use of immunotoxins in transplantation, cancer therapy and immunoregulation. *Transplantation*, **37**, 535–8.

Vitetta, E.S., Cushley, W. & Uhr, J.W. (1983). Synergy of ricin A chain-containing immunotoxins and ricin B chain-containing immunotoxins in *in vivo* killing of neoplastic human B cells. *Proceedings of the National Academy of Sciences, USA*, **80**, 6332–5.

Vitetta, E.S., Fulton, R.J. & Uhr, J.W. (1984). Cytotoxicity of a cell reactive immunotoxin containing ricin A chain is potentiated by an anti-immunotoxin containing ricin B chain. *The Journal of Experimental Medicine*, **160**, 341–6.

Vodkin, L.O., Rhodes, P.R. & Goldberg, R.B. (1983). A lectin gene insertion has the structural features of a transposable element. *Cell*, **34**, 1023–31.

Wada, S., Pallansch, M.J. & Liener, I.E. (1958). Chemical composition and end groups of the soybean hemagglutinin. *The Journal of Biological Chemistry*, **233**, 395–400.

Waksman, A., Hubert, P., Cremel, G., Rendon, A. & Burgun, C. (1980). Translocation of proteins through biological membranes. A critical review. *Biochimica et Biophysica Acta*, **604**, 249–96.

Waldmann, T.A. & Broder, S. (1977). Suppressor cells in the regulation of the immune response. *Progress in Clinical Immunology*, **3**, 155–99.

Walker, W.A. (1982). Mechanism of antigen handling by the gut. In *Clinics in Immunology and Applied Allergy* (Brostoff, J. & Challacombe, S.J. eds), vol. 2, pp. 15–40. W.B. Saunders Co. Ltd, London, Philadelphia, Toronto.

Walter, P., Gilmore, R. & Blobel, G. (1984). Protein translocation across the endoplasmic reticulum. *Cell*, **38**, 5–8.

Wang, J.L. & Edelman, G.M. (1978). Binding and functional properties of concanavalin A and its derivatives. I. Monovalent, divalent and tetravalent derivatives stable at physiological pH. *The Journal of Biological Chemistry*, **253**, 3000–7.

Wang, J.L., Cunningham, B.A. & Edelman, G.M. (1971). Unusual fragments in the subunit structure of concanavalin A. *Proceedings of the National Academy of Sciences, USA*, **68**, 1130–4.

Wang, P., Toyoshima, S. & Osawa, T. (1988). Concanavalin A-induced translocation of part of the GTP-binding activity from the membrane to the cytosol in murine thymocytes *The Journal of Biochemistry*, **104**, 169–72.

Wang, P., Matsumoto, N., Toyoshima, S. & Osawa, T. (1989). Concanavalin A receptor(s) possibly interact with at least two kinds of GTP-binding proteins in murine thymocytes. *The Journal of Biochemistry*, **105**, 4–9.

Warrington, R.J., Sunder, P.J., Olivier, S.L., Rutherford, W.J. & Wilkins, J.A. (1983). A human T cell-derived soluble factor able to suppress pokeweed mitogen-induced immunoglobulin production. *The Journal of Immunology*, **130**, 237–41.

Wassef, N.M., Richardson, B.G. & Alving, C.R. (1985). Specific binding of concanavalin A to free inositol and liposomes containing phosphatidylinositol. *Biochemical and Biophysical Research Communications*, **130**, 76–83.

Watkins, W.M. & Morgan, W.T.J. (1952). Neutralization of the anti-H agglutinin in eel serum by simple sugars. *Nature*, London, **169**, 825–6.

Wawrzynczak, E.J. & Thorpe, P.E. (1986). Enzymic removal of two oligosaccharide chains from ricin B-chain. *FEBS Letters*, **207**, 213–16.

Wawrzynczak, E.J., Falasca, A., Jeffery, W.A., Watson. G.J. & Thorpe, P.E. (1987). Identification of a tyrosine residue in the saccharide site of ricin B-chain using N-[^{14}C] acetylimidazole. *FEBS Letters*, **219**, 51–5.

Weaver, L.T. & Bailey, D.S. (1987). Effect of the lectin concanavalin A on the neonatal guinea pig gastrointestinal mucosa *in vivo*. *Journal of Pediatric Gastroenterology and Nutrition*, **6**, 445–53.

Weiel, J.E. & Hamilton, T.A. (1984). Quiescent lymphocytes express intracellular transferrin receptors. *Biochemical and Biophysical Research Communications*, **119**, 598–602.

West, G.B. (1983). Histamine releasers and rat mast cells. *International Archives of Allergy and Applied Immunology*, **72**, 284–6.

Wheelock, E.F. (1965). Interferon-like virus inhibitor induced in human leukocytes by phytohemagglutinin. *Science*, **149**, 310–11.

Whitfield, J.F., Durkin, J.P., Franks, D.J., Kleine, L.P., Raptis, L., Rixon, R.H., Sikorska, M. & Walker, P.R. (1987). Calcium cyclic AMP and protein kinase C-partners in mitogenesis. *Cancer Metastasis Reviews*, **5**, 205–50.

Wiley, R.G. & Stirpe, F. (1987). Neuronotoxicity of axonally transported toxic lectins, abrin, modeccin and volkensin in rat peripheral nervous system. *Neuropathology and Applied Neurobiology*, **13**, 39–53.

—(1988). Modeccin and volkensin but not abrin are effective suicide transport agents in rat CNS. *Brain Research*, **438**, 145–54.

Wiley, R.G., Blessing, W.W. & Reis, D.J. (1982). Suicide transport: Destruction of neurons by retrograde transport of ricin, abrin and modeccin. *Science*, **216**, 889–90.

Williams, P.E.V., Pusztai, A.J., MacDearmid, A. & Innes, G.M. (1984). The use of kidney beans (*Phaseolus vulgaris*) as protein supplements in diets for young rapidly growing beef steers. *Animal Feed Science and Technology*, **12**, 1–10.

Willie, A.H., Kerr, J.F.R. & Currie, A.R. (1980). Cell death: The significance of apoptosis. *International Review of Cytology*, **68**, 217–306.

Wilson, A.B., King, T.P., Clarke, E.M.W. & Pusztai, A. (1980). Kidney bean (*Phaseolus vulgaris*) lectin-induced lesions in the small intestine. II. Microbiological studies. *Journal of Comparative Pathology*, **90**, 597–602.

Wolff, C.H. (1987). Kinetics of Ca^{2+} uptake into lectin-induced secondary lymphocytes during reactivation with concanavalin A or interleukin 2. *Scandinavian Journal of Immunology*, **26**, 7–10.

Woodley, J.E. & Naisbett, B. (1988). The potential of lectins for delaying the intestinal transit of drugs. *Proceedings of International Symposium of Controlled Release of Bioactive Materials*, **15**, no. 73, 125–6. (Controlled Release Society Inc.)

Wouterlood, F.G., Bol, J.G.J. & Steinbush, H.W.M. (1987). Double-label immunocytochemistry: Combination of auterograde neuroanatomical tracing with *Phaseolus vulgaris* leucoagglutinin and enzyme immunocytochemistry of target neurons. *The Journal of Histochemistry and Cytochemistry*, **35**, 817–23.

Wright, C.S. (1980*a*). Location of the *N*-acetyl-D-neuraminic acid binding site in wheat germ agglutinin. A crystallographic study at 2.8 Å resolution. *The Journal of Molecular Biology*, **139**, 53–60.

—(1980*b*). Crystallographic elucidation of the saccharide binding mode in wheat germ agglutinin and its biological significance. *The Journal of Molecular Biology*, **141**, 267–91.

—(1981). Histidine determination in wheat germ agglutinin isolectin by X-ray diffraction analysis. *The Journal of Molecular Biology*, **145**, 453–61.

—(1984). Structural comparison of two distinct sugar binding sites in wheat germ agglutinin isolectin II. *The Journal of Molecular Biology*, **178**, 91–104.

Wright, C.S. & Kahane, I. (1987). Preliminary X-ray diffraction results on co-crystals of wheat germ agglutinin with a sialoglycopeptide from the red cell receptor glycophorin. *The Journal of Molecular Biology*, **94**, 353–5.

Wright, C.S. & Olafsdottir, S. (1986). Structural differences in the two major wheat germ agglutinin isolectins. *The Journal of Biological Chemistry*, **261**, 7191–5.

Wright, C.S., Gavilanes, F. & Peterson, D.L. (1984). Primary structure of wheat germ agglutinin isolectin 2. Peptide order deduced from X-ray structure. *Biochemistry*, **23**, 280–7.

Wright, H.T., Brooks, D.M. & Wright, C.S. (1985). Evolution of the multidomain protein wheat germ agglutinin. *The Journal of Molecular Evolution*, **21**, 133–8.

Yagisawa, H., Yamashita, Y., Yamagishi, S. & Sugiyama, H. (1985). Interactions of brain muscarinic acetylcholine receptors with plant lectins. *The Journal of Biochemistry*, **98**, 705–11.

Yahara, I. & Edelman, G.M. (1975). Modulation of lymphocyte receptor mobility by locally bound concanavalin A. *Proceedings of the National Academy of Sciences, USA*, **72**, 1579–83.

Yamamoto, T., Iwasaki, Y., Konno, H. & Kudo, H. (1985). Primary degeneration of motor neurons by toxic lectins conveyed from the peripheral nerve. *The Journal of the Neurological Sciences*, **70**, 327–37.

Yang, P., Baylin, S.B. & Luk, G.D. (1984). Polyamines and intestinal growth: absolute requirement for ODC activity in adaptation during lactation. *The American Journal of Physiology*, **247**, G553–7.

Yoneda, Y., Imamoto-Sonabe, N., Yamaizumi, M. & Uchida, T. (1987). Reversible inhibition of protein import into the nucleus by wheat germ agglutinin injected into cultured cells. *Experimental Cell Research*, **173**, 586–95.

Yoshitake, S., Funatsu, G. & Funatsu, M. (1978). Biochemical studies on ricin. Part XXII. Isolation and sequences of peptic peptides, and the complete sequence of Ile chain of ricin D. *Agricultural and Biological Chemistry*, **42**, 1267–74.

Younkin, S.G., Brett, R.S., Davey, B. & Younkin, L.H. (1978). Substances moved by axonal transport and released by nerve stimulation have an innervation like effect on muscle. *Science*, **200**, 1292–5.

Youle, R.J. & Neville, D.M. (1980). Anti-Thy 1.2 monoclonal antibody linked to ricin is a potent cell type specific toxin. *Proceedings of the National Academy of Sciences, USA*, **72**, 5483–6.

—(1982). Kinetics of protein synthesis inactivation by ricin anti-Thy 1.1 monoclonal antibody hybrids. Role of ricin B subunit demonstrated by reconstitution. *The Journal of Biological Chemistry*, **257**, 1598–635.

Zhao, J.M. & London, E. (1988). Conformation and membrane interactions of diphtheria toxin fragment A. *The Journal of Biochemistry*, **263**, 15369–77.

Glossary

Å: Angstrom

adenosine[4324]-nucleoside: one particular sensitive nucleoside of DNA

asialo-: de-sialylated

$\alpha_2\beta_2$: the subunit structure is usually given by denoting the subunit with a greek letter and the number of them in the molecule is given by subscripts

cAMP: cyclic adenosine monophosphate

CCK: cholecystokinine

CRM: cross-reacting material

DAG: diacylglycerol

DFMO: α-difluoromethylornithine

E & L-PHA: erythroagglutinating and lymphoagglutinating PHA, respectively

Fuc: fucose

G proteins: a family of intramembrane proteins linking external signals to intracellular events

Gal: galactose

GalNac: N-acetylgalactosamine

GALT: gut-associated lymphoid tissue

Glc: Glucose

GlcNac: N-acetylglucosamine

GNA: agglutinin from snowdrop (*Galanthus nivalis*) bulbs

GTP: guanosine triphosphate

GTPγS: a stable GTP derivative

HLA: histocompatibility antigens

IL-2: interleukin 2

IL-2R: receptor for interleukin

IP$_3$: inositol 1,4,5-triphosphate

Le^+ and Le^-:   Lewis positive or negative human blood groups
Man: mannose
M cells: microfold cells, gut Peyer's patch epithelial cells without microvilli
M_r: average molecular weight
mRNA: messenger RNA
NAc: *N*-acetyl
NTP-ase: nucleoside triphosphatase
ODC: ornithine decarboxylase
PHA: *Phaseolus vulgaris* lectin
pI: isoelectric point
PI-PLC: phosphoinositide-specific phosphodiesterase
PIP_2:   phosphatidyl inositol 4,5-diphosphate
PKC: phospholipid-dependent protein kinase C
PMA: 12-myristate 13-acetate
poly(A)-:   polyadenosine-
RCA: *Ricinus communis* agglutinin
SAMDC: S-adenosylmethionine decarboxylase
SAT: spermidine/spermine ^1N-acetyltransferase
SBA: soyabean agglutinin
SDS: Sodium dodecylsulphate
TPA: 12-*o*-tetradecanoyl-phorbol-13-acetate
transposon: transposable element in DNA
WGA: wheatgerm agglutinin

Index

gorse, *see Ulex*
G-proteins 80–2, 94, 172
Gramineae 9, 39
granulocytes 89
Graves' disease 95
Griffonia simplicifolia 13–14, 102
ground elder, *see Aegopodium*
growth 39, 42, 47, 50, 60, 62–3, 68–9, 85–
6, 111–13, 120–41, 158–65, 168, 171–
3, 178–9, 186–7
growth factors 90, 97, 119, 131, 140–1,
160, 178, 190–7
growth hormone 106
guanine 80
guar gum 20–3, 57
guinea pig 8, 77, 95, 167
gut 62–3, 103–4, 107–83
    endocrine cells 113, 134–44
    immune system 144–50
    lumen 109, 114, 124, 126–7, 132–42,
    145, 149–50, 160, 165–7

haemagglutination, *see* agglutination
haemagglutinin, *see* lectin
haemolysis 95
hamster 26, 105
hapten 3–4, 12, 58–9, 65, 69, 71, 78, 118,
147, 166
harmful effect, *see* damage
heat shock 71
heating 111–12, 172–3
Helminthosporium sacchari 62
helper cells 97–8, 104
hepatocytes 16, 78
hepatoma 92, 106, 195
hexokinase 89
histamine 103–4
histidine 11, 36, 102
histones 90, 96
homologies 32–5, 38–9, 44–5, 65, 69–70
hormones 2, 42, 105, 107, 136–44, 158,
179, 182
hormone mimicking effect 136
horse gram, *see Dolichos biflorus*
hostplant recognition 45, 64–70
host–symbiont relationship 64–70
human 15, 22, 26, 48, 76, 78, 82–3, 86–9,
95–9, 107–9, 112, 151, 164, 169–73,
190, 193, 204
Hura crepitans 49–50
hyacinth bean, *see Dolichos lablab*
hybridoma 89
hydrolases 118–19, 173

hydrophobic sites 36–7
hydroxyapatite 151
3-hydroxybutyrate 136
hydroxyflavones 64
hydroxyproline 9–10, 43
hyperplasia 113, 120–2, 124–5, 131, 135,
160, 172, 178–9
hypersensitivity, *see* allergy
hypertrophy 113, 120, 132, 140–1, 161
hypha 60

IgA 113, 144, 147, 149
IgE 102–4, 113, 144, 150, 204
IgG 25, 144, 147–9, 157–8, 204
ileum 125, 150
imbibition 41, 58, 60–1
immunity, immunology 1, 25, 59, 74, 85,
87–9, 95, 97, 99–100, 103, 114–16,
140, 144–50, 158, 169–70, 179, 182,
204
immunoglobulins 2, 10, 157–8
immunological methods 20, 39–42, 44–7,
49, 63, 68, 104, 108, 115–18, 134, 143,
156–8, 166–8, 184–8, 202
immunosuppression 88–9, 95–101, 104, 147
immunotoxins 180, 199–204
indomethacine 94
infants 160
infection 46–7, 61, 64–70, 150
    thread 66
inhibition, inhibitor 2–3, 6, 10, 12, 14, 19–
20, 23, 28, 31, 45, 47–8, 57, 60, 62–3,
66, 70–1, 80–1, 84, 87, 89, 91–2, 97,
99, 101, 103, 106, 113, 118, 120, 122,
124, 151–2, 155, 159–61, 163, 166,
170, 173, 178–9, 186, 190–3, 197
inositol and its derivatives 8, 36, 82, 84–5,
88, 171, 183
insects 62–3
insulin 90–1, 105–8, 140–4, 190–6
interference, *see* inhibition
interferon 49–89
interleukin 79, 83–100
internalization, *see* endocytosis
intestinal loop 166
intolerance 160
invertase 118
ionophore 49, 89
isoelectric focusing 20, 23, 25–6, 29
isolation, *see* purification
isolectins 11, 20–8, 33, 40, 46, 49, 53, 63,
76, 111, 142, 145–6, 169
isoproterenol 81